Ripples
on a cosmic sea

FRONTIERS OF SCIENCE
Series editor: Paul Davies

Ripples on a Cosmic Sea: The Search for Gravitational Waves
by David Blair and Geoff McNamara

Cosmic Bullets: High Energy Particles in Astrophysics
by Roger Clay and Bruce Dawson

Forthcoming:

Patterns in the Sand: Computers, Complexity, and Life
by Terry Bossomaier and David Green

The Feynman Processor: Quantum Entanglement and the Computing Revolution
by Gerard J. Milburn

Lamarck's Signature: How Retrogenes Are Changing Darwin's Natural Selection Paradigm
by Edward J. Steele, Robyn A. Lindley, and Robert V. Blanden

Beginnings of Life: The Origins of Life on Earth and Mars
by Malcolm Walter

Ripples
on a cosmic sea

The Search for Gravitational Waves

David Blair and Geoff McNamara

Foreword by Paul Davies

Helix Books

Addison-Wesley
Reading, Massachusetts

0-201-36082-9

A CIP record for this book is available from the Library of Congress.

Printed in the United States of America.

Published in Australia by Allen & Unwin Pty Ltd

Addison-Wesley is an imprint of Addison Wesley Longman, Inc.

Jacket design by Suzanne Heiser

123456789-DOH-0201009998
First printing, March 1998

Find Helix Books on the World Wide Web at
http://www.aw.com/gb/

To our children

Palenque, who grew up with gravitational waves, Carl and Linden, who always ask why and Amy Laura, who came into the Universe while the story was being told.

Contents

Series editor's foreword ix
Acknowledgements xiv
Prologue: 'Space is curved, you know' xv

Introduction: Why search for gravitational waves? xxvi

1 Space, lies, and geometry 1
2 Newton's space, Einstein's universe 11
3 A theory of some gravity 23
4 The cosmic looking-glass 38
5 Making waves 55
6 Stars that go bang in the night 65
7 The coming of the pulsars 79
8 Pulsars prove gravitational waves 95
9 Black holes and the beginning of time 104
10 The searchers 116
11 Super detectors 130
12 Shedding light on gravitational waves 147
13 New developments, future trends 163
14 The vision of gravitational wave astronomy 174

Epilogue: 'A gift of wonder' 177
Index 181

Foreword

Science, remarked the late chemist Sir Peter Medawar, is the art of the soluble. The frontiers of science are the boundaries between the do-able and the intractable, the known and the unknown. Those scientists working at the frontier seek to push back the boundaries of the do-able and reveal the unknown territory beyond. Because the frontier already represents the outer limits of scientific achievement, any further advance demands the utmost ingenuity and the use of every trick in the book. No subject better exemplifies Medawar's dictum than the search for gravitational waves. The detection of these elusive entities lies tantalisingly close to the edge of do-ability and the heroic attempt to turn that dream into reality is one of the most amazing scientific stories of our age.

The story began in 1915, with the publication of Albert Einstein's general theory of relativity, itself an intellectual feat of momentous proportions. General relativity, as it is normally called, serves both as a theory of space and time and as a theory of gravitation. Its predictions are literally cosmic in scope. An entire subject —cosmology—is founded upon it. Its more familiar predictions include the expanding universe, black holes and time travel. Its practical effects manifest themselves in areas as diverse as astrophysics and aircraft navigation.

Though most of the major predictions of general relativity have been carefully checked by experiment and observation, one key prediction of the theory remains to be fully tested. In 1918, Einstein himself noticed that the equations of general relativity possessed solutions describing wavelike undulations in the gravitational field. These 'gravitational waves' are not waves of any substance or medium, they correspond to ripples in spacetime itself, an eerie concept that never ceases to intrigue me. Like electromagnetic waves, discovered 60 years earlier by James Clerk Maxwell, gravitational waves move at the speed of light and carry energy. But there the resemblance ends. Whereas electromagnetic waves are readily harnessed—in the forms of laser beams or radio signals for example—gravitational waves are almost inconceivably weak. You could stand directly in a billion kilowatt beam of gravitational wave energy and not feel a thing. This extraordinary elusiveness makes gravitational waves virtually unobservable. And yet, according to general relativity, they should be there. For half a century after Einstein's discovery, gravitational waves were dismissed as a mathematical curiosity. Few people expected their effects ever to be observed. But one man, an American by the name of Joseph Weber, was captivated by a vision. If gravitational waves exist, he reasoned, it must be possible to find them by building a sensitive enough detector. The detector need to do no more than merely register the waves' fleeting passage. Given just the merest flutter of activity, it would be possible to test Einstein's longstanding prediction. But that would not be all. With a functioning gravitational wave antenna, mankind could also open up a whole new window on the universe.

Astronomy has long been dominated by the use of visible light as the main source of information about the cosmos. In recent years, the electromagnetic spectrum used by astronomers has been extended first into radio wavelengths, then into the infra-red, x-ray, ultra-violet and even gamma-ray regions. To study the universe using gravitational aves would not be an extension of this process, it would be the exploitation of an entirely new spectrum. What's more, it would be a spectrum capable of umveiling parts of the universe forever blocked to electromagnetic waves.

One of the most enduring memories as a young postdoctoral fellow in Cambridge in 1971 was listening to an upbeat lecture by the enthusiastic Weber. Here was a man truly fearless in the face of powers of ten. In his skilful experimental hands, they just dropped away like so much excess baggage. What began as a seemingly hopeless task, orders of magnitude beyond feasibility, ended up, after Weber had refined and refined the experimental process, to be—well, just about do-able. In fact, at that time, Weber thought that it was not only just about do-able, but just about done. He was cheerfully reporting tiny tremors in a pair of aluminium bars that he was convinced were caused by the disturbing effect of gravitational waves created in the centre of the galaxy. alas, further work failed to confirm this bold claim, but Weber did succeed in blazing a trail that many others were doggedly to follow.

Among those others was David Blair of the University of Western Australia. Sharing Weber's dream of opening up a new window on the cosmos, Blair sought to improve Weber's original trembling-bar design, using better sensors and a supercooled containment vessel. Then the field shifted direction with the realisation that laser beams bounced from distant mirrors might offer an easier method of detection and Blair began laboratory tests with a prototype system. Over the years, Blair became a world leader in this esoteric technology. He also became a tireless campaigner for government funds to back his project. With a punishing schedule of international meetings, he has sought collaboration with several other research groups in the northern hemisphere in an attempt to build a fully-functioning laser-based gravitational wave observatory in Australia.

The technology needed to achieve Weber's dream is truly mind-boggling. It demands detectors so sensitive they could pick up the equivalent of a pin being dropped on the other side of the planet. They must spot changes in length which, measure for measure, are the same as the width of a human hair in the distance to a nearby star. The vibrations they seek to monitor are so small that even the noise of atoms moving about drowns them out. This is the art of the soluble with a vengeance.

Can it really be done? Yes, it can. If a star is seen to explode in our galaxy tomorrow (such supernova events happen on a average every few decades), we would be able to pick up a burst of gravitational waves using current Weber-bar technology. But to make the project worthwhile, scientists want instruments sensitive enough to detect such events in faraway galaxies too, so they don't need to wait for years for something to happen. That means using laser interferometers. So, an already incredible level of sensitivity is being pushed still further. State-of-the-art laser systems are even now being built in the United States and Europe.

As the new millennium dawns, we can look forward to a new era in the exploration of the universe, seen through gravitational eyes. With a network of gravitational waves observatories around the world, it should be possible to record the death throes of massive stars as their cores implode, to follow neutron stars and black holes as they spiral together and plunge into each other and to probe the hearts of mysterious quasars—among the most distant objects in the universe. But, most important of all, the new astronomy may reveal phenomena in the cosmos that we never even suspected.

Whether Australia will play a central or merely peripheral role in this bold new venture remains uncertain. Blair's research group in Perth and other groups in Adelaide, Canberra and elsewhere, have the expertise to earn them a seat at the high table of gravitational wave astronomy. Their skills and knowledge have been hard-won over many years of persistent development work. What they lack is cash. Without adequate funding, they will mainly be relegated to the role of onlookers and a proud chapter in Australian science will be brought to a sad close.

There is no finer guide to this fascinating and challenging subject than David Blair. His deep grasp of both the theoretical and practical aspects of gravitational wave technology is second to none. He has personally made many of the delicate pieces of instrumentation needed to conquer those daunting powers of ten. Writing with author Geoff McNamara, Blair gives us a rare glimpse into the world of fundamental research, with the emphasis on the fun. The authors tell a thrilling story of hide-and-seek, of clever men and women pursuing a half-glimpsed treasure, confident in the expectation that the goal is within their grasp.

Paul Davies
Adelaide, July 1997

ACKNOWLEDGEMENTS

This book is the result of a long process. Paul Davies suggested it. Innumerable colleagues at conferences suffered my questions. My colleagues in the Australian International Gravitational Research Centre at the University of Western Australia put up with my absences as I escaped the laboratory to find time for writing.

Debbie Greenwood patiently suffered my handwriting. Then Geoff McNamara, whose science writing I already admired, agreed to take my technical compendium and massage it into a readable text. Finally, Ian Bowring and Emma Cotter at Allen & Unwin made it all happen, despite the frustrations involved in trying to pin down an illusive author. Thanks to all of you.

David Blair

Prologue

'Space is curved, you know'

I stood lost in thought on the beach, looking out at the calm ocean. It was a clear, still day and I could see as far as the horizon. Out there, I could just make out a ship bound for who knew where, ploughing towards the open sea. The sky was a brilliant shade of blue which reflected deeply in the calm water. Waves washed gently on the sand, leaving seaweed and foam. Each wave left its mark on the shore, its signature, only to be erased by a later wave. Sunlight reflected off the gently ebbing water, producing sparkles that resembled stars in a watery sky. How far away the horizon seemed on that warm, sunny day . . .

'It's curved, you know.' I turned, startled, at the voice. It was a man's voice, tinged with a foreign accent. Behind me an old man was standing. His untrimmed grey hair was swept back. His ragged moustache looked as though it had been there all his life. He must have walked up the sand while I was lost in thought.

The old man didn't look at me, but kept looking out to sea. 'Space,' he continued without invitation, 'it's curved, just like the ocean.' He turned towards me and gave me a look as if he knew me and expected me to know him. Who was this strange old man? What sort of person walks up to a complete stranger and simply starts making odd statements that deny common sense? How on earth could space be curved? Space is nothing, emptiness, and lots of it. In order for something to be curved, it has to exist. Space doesn't exist; it's simply . . . nothing.

The old man turned to me as if I was supposed to be moved by his assertion. I decided to humour him. 'How can you bend nothing?' I asked at last.

'Oh, it's possible,' said the old man, 'in fact it's difficult not to. Think about the ocean,' he said, gesturing towards the water. 'For centuries seafarers stayed close to the shore because they thought the ocean was flat. Since they could see an edge they assumed they'd fall off the ocean if they ventured too far. Now it's obvious that they were wrong. Even school-children know that if you venture far enough you'll eventually get right back to where you started,' the old man turned to me and pointed over his shoulder with his thumb, 'from the opposite direction!' He seemed to think this idea was an important one because he just stared at me seriously for a few seconds before continuing his spontaneous monologue. 'Space is the same!'

I had trouble with that. After all, space was just this three-dimensional nothing that we inhabit. You can walk around in it, drive through it, even map it. If I move up or down, that is one dimension. If I move left or right, that is the second dimension. If I move forward or backward, that is the third. If a plane was flying through the air, I could describe its motion through space by plotting its course and measuring its speed. Given enough information, I could even work out where it would be in a few hours from now. But the plane would be simply flying through space! The only reason it would end up back where it began—assuming it had enough fuel—is because it had followed a circle through space around the Earth. It certainly could have nothing to do with space being curved. Three dimensions equals space. I have three dimensions, and so does my house. So did this strange old man standing next to me on the shore . . .

'Are you trying to tell me that ship out there will end up on the opposite shore of this ocean because space is curved?' I asked.

'No, of course not.' He smiled, a gentle smile, not the kind that mocks your lack of understanding. 'It's just an analogy. A way of explaining what space is like. Imagine that you set out in a small boat and you rowed a few hundred metres off shore. Do you think you'd notice the curve of the ocean?'

'Well, no, I don't suppose I'd see that the ocean was anything but flat,' I admitted.

'And yet you know the ocean is curved,' he said.

'Yes, it's curved,' I replied.

He seemed satisfied with my answer. 'How far do you think you'd have to row before the houses behind us began to disappear below the horizon?'

'I'm not sure. Ten kilometres?' I offered lamely.

'I don't know either,' smiled the old man, 'and at any rate I think it would depend on the size of the houses and if there was much of a swell on the ocean at the time. But let's say you're right and after rowing ten kilometres out you begin to notice that the ocean is curved. When you were close to the shore the ocean seemed flat. It wasn't until you ventured far from the shore that you realised it's not that simple.

'Can you imagine how the early mariners must have felt as they ventured farther from the shore, when they saw their villages beginning to sink below the horizon?' I couldn't tell if he wanted an answer or not, but before I could think of what to say he answered for himself. 'They had no idea that the ocean was anything but perfectly flat. All their lives they had lived with, fished from and swum in an ocean that, except for the waves, had seemed to be perfectly flat. Now they began to realise that the ocean was, in fact, curved. How would you feel?'

I began to wonder where all this was leading. What had the ocean to do with space? There still seemed no connection between the ocean and space. For one thing, the ocean is a two-dimensional surface. It can be curved around the globe of the Earth every bit as easily as I can mould a sheet of wrapping plastic around an orange. Space—that three-dimensional nothingness we build cities in and which planets hurtle through—is a different matter. Oh, I'd heard of this 'curved space' business before: bending light rays, time changing. I'd seen *Star Trek*. But this old man seemed to think the idea was important, not just scientifically, but as if it could change our entire view of the world.

'OK, so you say space is curved like the ocean. How is it that when I look at the night sky everything looks normal? Surely if space were curved I'd see some evidence, perhaps the constellations constantly changing shape. It would be like looking at the sky through a fishbowl!'

'That's a very good point. What you have to remember is that when you look at the night sky you're still looking at the nearby cosmic shore. The fact is that even at the distance of the stars of our Galaxy, which stretch for a hundred thousand lightyears, you're still too close; you're like the boat only a few metres from the shore; you haven't ventured far enough to see your home disappear below the horizon. Yet. But travel far enough and you'll see exactly what I've been saying. Space is curved. And you'll feel just the way those early mariners did: perplexed . . . and just a little frightened.'

It seemed unlikely that humans could ever travel so far into space, at least in my lifetime. In the brief but glorious history of the space age, we've managed to send spacecraft to the farthest reaches of the solar system. The two Pioneer spacecraft had visited Jupiter and Saturn. The Voyagers had made it to Uranus and Neptune. All four were now flying out of the solar system, but even at their tremendous speeds and after decades of flight, they had yet to find the boundary where the Sun's influence merged with the distant stars. The difference between the distance to the farthest planets and the distance to the stars was like getting your toes wet compared with swimming to the horizon.

'Suppose you're right . . . '

'I am,' he interjected.

I was surprised by the old man's confidence. 'Isn't it unlikely we'll ever be able to prove that space is curved? I mean, there doesn't seem to be a way to travel to the nearest stars, let alone outside the Galaxy to see what you've been talking about.'

'That's quite true,' he nodded, 'it's unlikely we'll be able to travel much farther than the shores of the solar system—at least for the foreseeable future. But we can look out into the cosmic sea using telescopes.

'Imagine that your eyesight was poor,' he continued. 'Suppose you were so short-sighted that you couldn't see more than a few metres in front of you, and that everything beyond was a complete mystery. If there was a ship out there, you'd never know it. You could only guess what lay beyond and would never be able to see the horizon without help. Assuming you only know what your myopic eyes tell you, what would you know about the sea?'

I had to admit that without any information to the contrary, I'd have to assume the rest of the sea was as flat and two-dimensional as what I could see. 'I'd assume it was flat everywhere,' I said.

'That's just what Galileo and Newton thought, too,' he mused. 'But then telescopes began to improve. We began to see further than ever before. Giant telescopes on Earth's surface and in orbit have begun to probe the depths of space. Astronomers have discovered that the Universe does contain "fishbowls", as you put it. We see great and distant clusters of galaxies swimming and distorted in the sky. In reality it's their images that are being distorted by the varying curvatures of space.'

The old man seemed preoccupied by this thought. He simply looked out at the receding ship that had now reached the horizon. OK, so space might be curved. I didn't really understand it, yet, but the ocean analogy did seem to make sense.

'Is the entire Universe curved? Like a single fish-bowl?' I asked.

He turned back to me with an excited look in his eyes. 'Yes, and that's one of the most important points! On the largest scale imaginable—a scale that encompasses the entire Universe—space is curved. But there are local distortions as well, like the galaxy clusters I just described to you.'

At this point, he moved so close to the water I thought he'd get his old leather shoes wet. He didn't seem to mind. 'Come here,' he gestured. He crouched down on the sand and looked intently at the small waves washing towards his shoes.

'What do you think causes those waves?' he asked, turning up to look at me.

'The wind, distant storms, far-off boats, all sorts of things,' I replied, thinking it was a terribly obvious question.

'You're probably right,' he said, turning back to the water. 'But imagine that ship out there was ploughing through a dead calm sea, a sea so smooth that the tiniest movement sent waves in all directions. Can you imagine it?' he asked.

'Yes, I can,' I replied, 'They would move outward, always getting smaller. Eventually they might become the ripples here on the beach, but by then they might be very small.'

'Yes, perhaps they would be almost imperceptible and difficult to detect.' The old man seemed lost in thought again.

The old man suddenly stood up, saying, 'But come over here, I want to show you something.'

I followed him over to a rocky outcrop. He stopped by a rock pool filled from the last high tide. Stooping, he picked up a small shell and dropped it into the pool. I assumed this was just some idle action, but he suddenly grabbed my arm and said 'Watch! Watch!' For a moment I was confused, but then he pointed out what he wanted me to see. 'There, do you see them!' he said excitedly. 'The ripples! the ripples!' I could, indeed, see a procession of tiny ripples moving quickly across the pool, soon disappearing against the opposite edge. 'You can do that in space, too,' said the old man gazing at the last of the ripples as they faded from view.

'Are you trying to tell me that ripples can form in space just as they do in the water?' I asked.

'Absolutely,' was the reply. 'If you drop a big enough "shell" you can create ripples in the fabric of space. What's more important is that if you're sitting on the opposite side of the "pool" you can detect those ripples, maybe even see where they came from. It might even be possible to learn what sort of "shell" caused them in the first place. Wouldn't that be something?' He looked at me eagerly, but he soon became thoughtful again. 'The only problem is that space is very rigid,' he said enigmatically.

'Rigid?' I prompted, now hoping he would continue.

'Yes, rigid,' he continued. 'The old view of space was that it was infinitely stiff, a vast expanse of nothing, a rigid concept of three dimensions. Nothing could bend it, let alone produce ripples. We now know space is something very real, something capable of bending, but at the same time something very, very, stiff. In fact,' he said, turning back to the ocean and looking in the direction of the ship, now almost gone from view over the horizon, 'if the ocean were made of steel and that ship out there were on wheels, its ripples across the steel sea would be billions of times larger than any sort of wake it would leave in space.'

'What sort of object could produce ripples in space if it's that rigid?' I asked at last.

'Oh, there are lots of them,' he replied, 'collapsing stars, coalescing pairs of neutron stars, the formation of black holes, even the birth of the Universe itself is thought to have left its wake in space. The problem is that even these violent cosmic catastrophes produce only tiny ripples, because space is so stiff, you see.'

He seemed to read my reservations on the whole matter. 'Think about the ship out there. You said yourself that the ripples it produces would be incredibly difficult to see by the time they reach the shore. How much more so if the ocean were made of steel?'

'What's the point, then,' I said, 'if these ripples in space . . . '

'Gravitational waves. They're called gravitational waves,' he interrupted.

'These gravitational waves, then,' I continued, 'are so near impossible to detect, why bother looking for them?'

He suddenly looked very pleased with himself. 'You should have asked the same thing of Friedrich Hertz,' he grinned.

'Hertz?' I asked.

'Back when Hertz discovered radio waves, nobody had any idea of the implications of the discovery. Now we can't imagine life without radio waves. Everything depends on them: communication, navigation, defence, and of course a great deal of exploration of the Universe is carried out using radio waves.'

'Yes, but what has that to do with gravitational waves?' I persisted. 'After all, these so-called gravitational waves must surely be purely hypothetical, some fancy of a physicist's imagin——'

'No!' the old man suddenly exclaimed. 'They're not

theoretical. Oh, they were in my time, but not so long ago their existence was proved beyond the doubt of any reasonable person. Don't you see?' asked the old man turning to face me. His face glowed with enthusiasm. 'This could be the beginning of a whole new era in astronomy!' He spoke as if this was one of the most important times in the history of the world.

'Look,' he said, picking up another shell. 'If I throw this into the water, thus . . . ' splash, went the shell, ' . . . and you're standing on the other side of the pool, you can detect the falling shell simply by seeing the ripples. Not only that, but you can tell where the ripples came from and how big they are. Even if you could see nothing else, you'd see those ripples and know that something had happened somewhere else in the pool.'

I was beginning to understand what he was getting at. With nothing more than a fisherman's float, I could detect the impact of a shell into the water—a major event in a pool this size. If I were able to do the same thing for gravitational waves passing through space, I could listen in on some of the most violent events in the Universe . . . right back to the Big Bang itself! I began to see what the old man was excited about.

'So you're saying we could explore the Universe in an entirely new way?' I asked.

'Exactly.' The old man seemed more excited now that I was beginning to understand what he was on about. 'When physicists detect gravitational waves here on Earth, they'll have opened up an entirely new window on the Universe. Since antiquity humans have studied the heavens in visible light. The detection of radio waves from space allowed us to look at the cosmos in an entirely new light. Since then we've opened up even more parts of the electromagnetic spectrum—X-rays, gamma rays, ultraviolet and infrared—each new window revealing a Universe which is entirely different from what we knew before. But the detection of gravitational waves won't be simply the discovery of a new part of the electromagnetic spectrum like light or radio waves. This time it will be the opening of an entirely new spectrum, the spectrum of gravity!'

'But how do you detect gravitational waves?' I said.

'Ah, that's the rub,' said the old man. For the first time he seemed happy not to tell me something. 'It's not easy. Not at all easy. Space is very stiff, remember. But,' and he moved closer to me, speaking almost confidentially, 'they're at it already. Around the world physicists are setting up detectors and are patiently waiting for a shell to fall into the cosmic sea. When it happens, they'll be ready. They'll be ready . . .'

The old man turned and began to walk away. 'Hang on a minute,' I called after him, 'you haven't finished the story. Who are they? How are they searching for gravitational waves? And I still don't understand how you can have ripples of nothing.' Not to mention the fact that I still didn't know who the old man was.

The old man turned back. He pulled an old paperback out of his coat pocket and held it out to me. 'Here, read this.'

I took the book from the old man's hand. It looked as if it had been read a hundred times—its pages were worn and grubby, the cover coming away from the spine at the top and bottom. Here and there pages had been dog-eared as if to mark an unusually significant passage. The cover was worn and creased, but I could still read the title: *Ripples on a Cosmic Sea*, by David Blair and Geoff McNamara. On the back cover were a few words still legible, saying something about gravitational waves opening up a new window on the Universe.

By the time I looked up, the old man was again walking along the beach. I called out 'But you still haven't explained how nothing can be curved.'

'Read the book,' he called without turning back.

I thought about going after him, but something told

me he'd say nothing more on the matter. I looked down at the worn paperback. 'Read the book,' he had said. Would this explain how waves can travel through nothing? I went to call after the old man again, but he had disappeared.

The beach was now empty, save for the occasional gull wandering among the debris washed up on the shore. The ship had now disappeared over the horizon. The ripples in the pool had faded. But along the beach, waves from unknown sources lapped at the sand. Could the Earth be like that shore, constantly washed with gravitational waves? Were these waves really as important as the old man seemed to think? Were we like the early mariners about to set out on a new gigantic ocean?

The old man had teased me, given me some tasty morsel to think about. Now I had to know. I sat down on the rocks and began to read . . .

Introduction:
Why search for gravitational waves?

Gravitational waves are ripples in the fabric of space-time. They were predicted by Einstein and proved to exist by certain astronomical observations. However, they have not yet been directly detected. They carry away enormous amounts of energy from some of the most violent events in the Universe, and yet our technology will detect them only as the faintest of whispers. So sure are physicists of the existence of gravitational waves, however, that huge resources are being channelled into building large gravitational wave detectors around the world. Using these detectors, astronomers hope to listen to the Universe using this entirely new spectrum.

Whenever a new method of astronomical observation is developed, it leads to a new, improved understanding of the Universe. The invention of the telescope, for example, caused a major revision of models of the solar system. One of the major players in the early telescopic era was Galileo Galilei. In Galileo's time, the sixteenth century, the Earth was considered to be the centre of the Universe, in accordance with biblical teaching. While Copernicus and others challenged this idea, proposing instead that the Sun was the centre of the Universe, there were few observations to support either view. Galileo's telescope allowed him to see the moons of Jupiter, the craters, mountains and valleys on our own Moon, and the myriad of stars that make up the Milky Way. While

Galileo's observations didn't prove the falsity of the 'official' Earth-centred Universe, they did make scientists reconsider. Eventually, many decided that Copernican Sun-centred theory explained the observations of Galileo and others in a more satisfactory way.

Our knowledge of the Universe grows with every advance in telescope design and size. The Universe as we know it today is much richer and more complex than Galileo could have imagined. However, many people alive at the time of major discoveries have been unaware of the tremendous significance of the discoveries made in their time. Few non-scientists of Galileo's time, for example, understood the fantastic implications of his observations, and yet they represented a turning point in our thinking. In a similar way, few people are aware of the importance of the search for gravitational waves. That's why we have written this book: to explain to you, the non-scientist, one of the most important and exciting scientific adventures of our time.

Why is the search for gravitational waves so important? In the twentieth century it is tempting to think our understanding of the nature and origin of the Universe is almost complete. We can now see deeper and more clearly into the Universe than ever before. In recent years astronomers have glimpsed the very birth of galaxies and, through measurements of the Universe as a whole, are gradually redefining our knowledge of its age and its eventual fate. Despite these successes, plenty of problems remain. In particular, we are as far away as ever from understanding why the Universe is like it is. To solve such problems, we need new ways of looking at the Universe, a new angle on the problem. Even then, there can be no guarantee that the Universe will give up its secrets. But as every detective knows, a new line of enquiry, new clues and new evidence always help in securing a conviction.

Since Galileo, astronomers have been peering at the Universe through an increasing range of bands of the electromagnetic spectrum. Whenever a new part of the electromagnetic spectrum is opened, it is like opening a new window onto the Universe through which we see things totally unexpected. At first it was simply light which was harnessed and magnified through telescopes, which continue to grow in size and number. Later, radio waves were discovered and led to the discovery of a Universe far stranger than anyone could have believed. Still later came the development of X-ray telescopes and gamma ray telescopes on spacecraft, as well as ultraviolet and infrared telescopes. Gradually the electromagnetic spectrum has been extended and examined; every extension has led to dramatic new discoveries. Gravitational waves represent a completely new and unexplored spectrum, not a new window in the electromagnetic spectrum, but a truly new spectrum with which to explore the Universe. Whereas electromagnetic waves have given us an extended sense of sight, gravitational waves give us a new sense: the sense of hearing. Gravitational waves will allow us to listen to the Universe: we expect to be able to listen to the explosion of stars, the merging neutron stars, the creation of black holes, and even the sound of the Big Bang itself. Maybe with this new sense we will begin to understand why our Universe is as it is.

As with a newly discovered but unexplored continent, however, no one knows what lies ahead in the field of gravitational wave astronomy. It may be a wonderland of new discoveries, or it could be that this new land will be nothing but desert. We have to look, however. History is full of examples of important discoveries following, what seemed at the time, bizarre theories. Electromagnetic radiation was first predicted by James Clerk Maxwell in 1865, but not observed until 1886 by Heinrich Hertz. A century later, we cannot think of life without electromagnetic radiation—the entire electromagnetic spectrum has been harnessed for both astronomy and technology. The discovery of gravitational waves may follow a similar path. Fifty years after Maxwell, Einstein's revolutionary theory of general relativity predicted the spectrum of gravitational waves. For forty years physicists were not even convinced that Einstein's predicted gravitational waves would be detectable, no matter how good the technology became. Seventy years after Einstein's prediction, however, the existence of gravitational waves was proved by Joe Taylor through his observations of the behaviour of a strange stellar system called a binary pulsar. This discovery was recognised with the 1993 Nobel Prize for physics, to Taylor and his student Russell Hulse.

This is the story of the search for gravitational waves: the people, the concepts, the successes and the failures. The search for gravitational waves spans oceans and galaxies. It calls for the most sensitive detectors ever made, listening for the echoes from some of the most energetic and violent processes in the Universe. It involves the insight of a genius, unfulfilled predictions, chance discoveries and perseverance. The story will take us on a journey from the abstract to the practical. Along the way we'll see how stars are born, live and die; how a star thousands of times the size of the Earth can collapse into a sphere the diameter of a small city; how matter can disappear from this Universe forever, leaving behind a 'black hole' in space; and how these mysterious events can create gravitational waves. We'll see how space itself has physical properties that allow it to carry these waves. Finally, we'll peek inside the gravitational wave observatories of the late twentieth century, taking a look at some of the high technology and high hopes involved in the race to detect gravitational waves.

It's an unfinished story, however: no one has yet detected gravitational waves and it is impossible to predict when they will. It could be before this book is published, or it may not be for decades. One thing is certain, however: scientists will detect gravitational waves simply because it is a matter of detecting a phenomenon whose effects have already been observed in nature. The detection of gravitational waves simply depends on the development of the detectors with sufficient sensitivity. In that sense, it's a race of technology as much as science.

Around the world, gravitational wave researchers are creating the ears with which we will be able to hear the Universe. The new spectrum will give us a new sense. When astronomers finally detect gravitational waves, we will at last be able to hear the sounds of ripples on a cosmic sea lapping on our Earthly shore.

Chapter 1
Space, Lies, and Geometry

What is space? Not just the emptiness between the stars and planets, not just outer space or inner space. Not just the emptiness between objects, but the space that the objects occupy. We are so used to space, to living in space and moving through it, that it's difficult to recognise any sense in the question 'What is space?'

Modern physics has unified space and time into the concept of four-dimensional space-time. But what is time? Is it something that carries us along like a river, or is it more like a freeway we travel along? Whatever it is, most of us feel it goes too quickly between birthdays and too slowly during traffic jams. While we're quick to complain about time, do we really know what it is? It's funny how, on a long journey, we complain about the slowness of time but not about the bigness of space. Yet both are linked. That's why space-time is a useful concept.

So, is space just this nothingness in which all the objects of the material world exist? If it is, is time just a nothing in which all events occur? Definitely not! We want to show you that space-time is stuff with real properties. In particular, it is elastic. Understanding the elasticity of space-time allows us to understand gravity and gravitational waves. No one really knows why space-time is elastic, yet it is because of this elasticity that the Universe is a stranger and more interesting place than most people realise. The elastic property of space-time allows gravitational waves to ripple through the Universe at the speed of light and provides us with a whole new spectrum awaiting exploration—the spectrum of gravitational waves.

We grow up with 'common sense' ideas of the nature of space and time. In school mathematics, we are taught about Cartesian co-ordinates—x-axes and y-axes and rectangular grids of perpendicular lines in which we measure the paths and positions of objects. We use these concepts every time we look at a street directory. In geometry, we learn how the sum of the angles of a triangle add to 180 degrees and that parallel lines go on forever without ever intersecting—theorems first proved by Euclid. In physics we were taught how to add and subtract velocities, usually called Galilean or Newtonian relativity. We learnt either formally or informally that if you run at 10 kilometres per hour down the corridor of a train travelling at 60 kilometres per hour, then your speed relative to the ground outside is either 50 kilometres per hour or 70 kilometres per hour, depending on which direction you are running.

Having been taught the 2200-year-old geometry of Euclid and the 300-year-old physics of Galileo and Newton, it is not surprising that modern physics seems confusing to many people. Most of us are never told that these teachings are actually wrong—deeply and fundamentally wrong. They are founded on incorrect assumptions and create mental blinkers which make modern physics seem weird and incomprehensible. The problem is that these untruths actually appear to be correct when applied to everyday experience. This is a problem, because these incorrect assumptions present a barrier when we try to understand the nature of the large-scale Universe.

Ptolemy taught that the Earth was the centre of the Universe, that the Sun revolved around the Earth, and that the planets revolved around the Earth in complex epicycles—orbits within orbits. The theory worked reasonably well, and allowed you to predict the movements of the Sun, moon and planets. You could use the theory, for example, to predict the time of year when Venus would be an evening star or a morning star, or the passage of Mars and Jupiter through the various constellations.

The fact that the theory works for such predictions did not mean the Earth was at the centre of the Universe, however. Copernicus offered an alternative to Ptolemy's theory which was much simpler and much more accurate, but which relegated the Earth to being a humble planet along with all the rest. Over the centuries, evidence proved Copernicus's view, and today it is inconceivable that a school teacher would present Ptolemy's theory in any other than an historical context.

In a similar way, our knowledge of the nature of space has changed over the years. Gauss in the 1820s was the first to question the truth of Euclidean geometry. He refused to accept Euclidean geometry at face value, and realised that space need not have the geometrical properties of flat paper. In the 1850s, Gauss's student Reimann developed the geometry of curved space and showed that Euclidean geometry was just a special case of a much more general geometry. As we will see later on, you could now relate the properties of, say, triangles on surfaces to the overall shape of the surface, such as a sphere, a cylinder or a saddle.

Later in this chapter we will see how two American physicists, Albert Michelson and Edward Morley, showed that Galilean relativity is totally wrong when applied to light. They showed that when you shine a light beam down a moving railway carriage, the speed of the light beam relative to the ground outside is totally independent of the speed of the train. In 1906 Einstein went further by showing that Galilean relativity breaks down not only for light but for all moving objects. In general, you cannot add and subtract velocities in the simple Newtonian way. When you run down the carriage at 10 kilometres per hour, calculating a total speed of 70 kilometres per hour is a good approximation, but it is not exactly true.

Finally, in 1916, Einstein showed that Euclidean geometry cannot be applied in any place where there is gravity, and gravity acts everywhere in the Universe. In other words, Euclidean geometry is an incorrect description of geometry in our Universe. The sum of the angles of a triangle is not 180 degrees; lines which seem parallel do not go on forever without intersecting. Gauss's scepticism was correct. Euclid was wrong. Space is curved! Gravity is simply a manifestation of curved space.

If all this physics and mathematics is wrong, why do we go on teaching it at school? We teach the truth about the solar system; we teach the truth about atoms and molecules; we teach the truth about biology and the evolution of species. So why do we continue to teach lies about space? Call them 'incorrect theories' if you like, but the point is that these teachings are known to be incorrect. Why do we continue to teach these concepts in an age when there is a broad understanding that they are untrue? The fact is that they are a good approximation in our everyday world; space on Earth is very nearly flat, Euclidean geometry is very nearly true, as is Galilean relativity. Conversely, the unfamiliarity of the true nature of space makes it seem strange and unreal. Yet understanding curved space is as easy as understanding triangles drawn on the surface of an orange or a balloon, as we will see in Chapter 3.

The geometry of curved space can be illustrated by an analogy with the layout of roads over a hilly landscape. There is a superb example of this near Buenos Aires airport in Argentina. Viewed from an aeroplane, the roads appear to be laid out in a rectangular grid. Except something is wrong: they meander, sometimes closer together and sometimes farther apart. Is this haphazard Argentinean surveying? Despite their curved appearance from above, the streets are locally straight. What you see is a grid of streets built over a hilly landscape. The surveyors had two choices. They could create a grid of imaginary lines above the landscape and project a rectangular grid pattern straight below. Or they could stay on the ground and line up their poles one after the other up and down the hills. This is the easiest way, and this is how they did it in Argentina. The surveyors' poles were all correctly lined up, but the space was curved. A car with its steering locked straight forward would in principle follow the roads perfectly, in spite of the fact that viewed from far above the roads appear to meander. In the same way, the shortest line between two points on the surface of the Earth is a curve called a geodesic. Aircraft follow these trajectories. They are straight lines in the curved two-dimensional surface of the Earth.

Similarly, artists are able to convey the illusion of curved surfaces—three-dimensional images of people, houses, landscapes and still-life—on the flat surfaces of their canvases. The point is that our minds are quite capable of interpreting the geometrical properties of curved surfaces. The concept of the curvature of space is a relatively simple extension of this. What makes it seem strange is a lack of familiarity with this perfectly reasonable concept.

Euclidean geometry is not a very good approximation in everyday life. True, it works well for lines drawn on a flat piece of paper, but that's where its usefulness ends. Maps of the world that ridiculously enlarge Greenland and Antarctica are a result of the impossible conflict of transforming the spherical geometry of the Earth to flat paper. The construction of roads across a hilly landscape calls for more than the restrictive Euclidean geometry.

And yet when our formal learning of geometry is locked into flat pieces of paper, it restricts our minds enormously. The result is that when someone like Einstein suggests that space is really curved, we shrink back in horror: how could it be? The idea is absurd! Only a mathematical genius could understand such a concept! But it is no harder to come to terms with the concept of curved space than it was to abandon Ptolemy's Earth-centred Universe, or to embrace the idea of the evolution of species.

Einstein taught us not only that space is curved, but also that there can be ripples in that curvature. That is, the sum of the angles of a triangle differs from 180 degrees and also varies with time. Einstein's theory also tells us that, although space curves and bends, it is the stiffest, most rigid stuff in the Universe—while flexible, space is very difficult to bend. So space is a very tangible material with tangible properties like stiffness and shape.

But how can space carry waves? In the nineteenth century it was generally assumed that space was filled with some material. Scientists gave a name to this material: the 'aether'. It seemed necessary to have such a medium to explain how light waves could travel through the vacuum of space. This is easy to understand when we compare light with other types of waves, such as water waves or sound waves. Both of these forms of waves are carried by a medium—water or air. In a similar way, it's easy to see a wave travel down a long piece of rope or an uncoiled electrical lead. When you give the lead a flick with your wrist you can watch the wave travel from one end to the other. The aether, it was thought, was the medium that allowed the passage of light waves.

The aether was invented to allow light to travel across the emptiness of space. While sound needs some medium to travel through, such as air, it was known that light waves do not need air to sustain them. If you put a torch in a vacuum tank and suck out all the air the torch does not suddenly go dark. Similarly, light reaches us from the Sun and the stars across the vacuum of space. The aether, on the other hand, could not be sucked out using a vacuum pump. It permeated everything. Yet there was a simple test for its existence, which is easy to understand if we think about water waves and sound waves.

In a fast aeroplane, such as the Concorde, you can exceed the speed of sound. When this happens, no sound created behind you can ever catch up with you. If you travel in a boat towards oncoming water waves, the waves appear to approach you quickly.

As you ride over these waves, they appear steep and frequent. But if you turn the boat around you can travel in the same direction and at the same speed as the wave so that, relative to you, it appears not to be moving at all. If you sailed just a little slower than the waves, you would feel your boat gently rise and fall as the waves passed beneath you. At a very special sailing speed, the wave speed appears to be zero and the wavelength infinite. You would continually sail down the wave which seems to go on forever. This is exactly the principle of surfing.

The important thing is that the speed of the wave is tied to the medium. The speed of the wave is relative to the speed of the medium, and it makes no difference if you move relative to the medium or if the medium moves relative to you. In 1887, the two American physicists mentioned earlier, Albert Michelson and Edward Morley, set out to do the same thing for light. (Michelson actually tried the experiment first in 1881.) If there was an aether, then the Earth would be travelling through it as it orbits the Sun. In addition, the Earth rotates at about 1600 kilometres per hour at the equator. If light was travelling through the aether the way waves cross the ocean, then the speed of light travelling through the aether should appear different at different times of day. Looking at the sky overhead at dawn, for example, you are facing in the direction of the Earth's travel through space, 'facing into the wind' if you like. At sunset, you are in the hemisphere that faces in the opposite direction of the Earth's motion through space. Above you at sunset, the Earth's invisible wake is before you. If the relative motion of the aether varies through the day due to the rotation of the Earth, then so should the speed of light.

Michelson and Morley didn't measure the velocity of light directly, but compared the relative velocity of two light beams at right angles to one another. They did this by splitting a beam of light with a half-silvered mirror called a beam-splitter. Half the original beam was sent down one path, while the other half was sent down a path perpendicular to the first. The beams were then reflected back on themselves and recombined.

Michelson and Morley's experiment is fundamental to our story. Not only did it uncover some of the peculiarities of space, but it also turns out that the same device in a highly advanced form is an ideal gravitational wave detector. Remembering that light is a wave, we can draw the beams as oscillating wave forms. The outgoing waves of Michelson and Morley's experiment were in phase with each other because the paths are exactly equal. Such waves will add up to make a single, stronger wave. An observer would see a bright light.

However, if one part of the wave is delayed by half a wavelength compared with the other, the waves will exactly cancel. A rising crest will superimpose on a falling crest, creating a null: an observer would see no light. This phenomenon is called interference. The device which allows the waves to add or cancel is called an interferometer.

Michelson devised his interferometer with the financial help of Bell, who had invented the telephone five years earlier, in 1876. For a real interferometer there can easily be millions of cycles, or wavelengths, along each light path. Green light, for example, has about two million wave crests per metre. So a change of only half a wavelength in a few million wavelengths takes you from dark to bright. Clearly the device is very sensitive.

Now, suppose one of the beams is travelling upstream in the aether due to the motion of the Earth; then the second beam, perpendicular to the first, is sideways on, and should be mostly unaffected. The light on the upstream track will go more slowly, while the light on the return path will speed up. You might expect that these opposing effects would cancel out, but they can't! The sum of the higher speed and the lower speed gives you a net delay. This may be hard to believe, but it is easily proved.

To convince yourself, try a simple experiment next time you come across a moving walkway of the type you find at airports. To complete this experiment successfully, you must be brave enough to suffer a few indignant looks! Wait until the walkway is empty, then take a brisk walk down the walkway, but before you reach the end turn around and walk back to the start. Repeat the experiment beside the walkway, timing yourself each time. The return trip on the walkway will always be slower. In the worst case, your walking speed exactly equals that of the walkway. Then the upstream journey takes infinitely long. Despite your rapid speed in one direction, the total journey time is stretched out indefinitely. Michelson and Morley used this idea to search for signs of the upstream–downstream effect.

Despite the sensitivity of their experiment, they found nothing. There was no discernible difference in the velocity of light due to the motion of the Earth or anything else. This result was astounding: it meant that the speed of light never varied, no matter how the source or the measuring apparatus moved. Hence, there could be no aether. In most people's eyes, space reverted to being an empty conceptual grid. Light did not need a medium to sustain its oscillations. Space was simply nothing.

Einstein used the universality of the speed of light in his special theory of relativity in 1905. This theory ignored gravity, but reconciled the universality of the speed of light with everyday properties of water waves and moving objects. But space was still an empty conceptual grid.

This concept of space was to remain unchallenged until 1915, when Einstein shocked the scientific community and perplexed the rest of the world with a new description of gravity and space. In Einstein's Universe, not only was space real, but it was curved, and it was awash with waves like a cosmic sea.

CHAPTER 2
NEWTON'S SPACE, EINSTEIN'S UNIVERSE

Like many aspects of nature, gravity is one of those things that most of us take for granted. It's just this force that keeps us on the ground, that makes it hard to climb up hills, and causes leaves to fall. The same goes for space and time: space is simply an empty volume filled with the various bits and pieces that constitute the natural Universe. The bending of space, let alone its ability to carry waves, seems at best counter-intuitive, at worst nonsensical. We're almost too familiar with gravity, so when scientists come along and tell us what gravity and space are really like, these everyday concepts suddenly become abstract and unrealistic.

We've seen how the old ways of thinking about space are incorrect. Despite the fact that the centenary of general relativity is now within sight, the old, false ideas of the nature of space persist. Nonetheless, to appreciate the nature of gravitational waves, it's necessary to understand the true nature of gravity and space. Rather than diving straight into general relativity, we'll start by reviewing the physics of Newton, whose genius allowed the discovery of new worlds, and whose physics is at least intuitively correct. But it's the Universe revealed to us by Einstein that allows the passage of gravitational waves. So, standing on Newton's shoulders, we'll peek into Einstein's Universe. From this high vantage point, we will be able to see ripples in the fabric of space-time.

Isaac Newton was an unpleasant genius who followed hot on the heels of Galileo Galilei. He put Galileo's words on forces and motion, gravity and planets to the music of mathematics. Newton took commonly experienced physical forces and defined them. We all experience physical forces every day. Take doing the shopping, for example. We have to keep a tight rein on the shopping trolley in most supermarkets, not only because of the castors that seem to have minds of their own—all four of them seem to be going in different directions at once—but when the trolley is full you have to pull and push pretty hard, sometimes both at once, to get the beast to the checkout. This may seem like one of life's chores but it is, in fact, a wonderful experiment in physics. What the trolley is saying to you, other than the wheels need replacing, is that it has mass and that once that mass has motion you have to apply a different force to change its direction. When the trolley is empty, it has less mass and requires less effort on your part to move it or slow it down. When it's full, you'll notice that it's harder to stop and start again because it has more mass. If you stop it while you make a last-minute check of the weekly specials, you'll find that it's harder to get going again. This time it's saying to you that it's quite happy to stay where it is unless you apply force, that is, give it a decent shove.

Newton saw the same thing, although we doubt he ever went shopping this way. Newton defined these common experiences in a set of wonderful laws of motion. And here they are:

1. A stationary object remains stationary unless a force acts on it (the trolley stays put until you push it).
2. A moving object continues to move in a straight line unless forces act on it (the trolley collides with an indignant shopper).

3. If forces act on an object, then it accelerates in proportion to its mass (a heavy trolley is harder to push or slow down than an empty trolley).
4. Whenever a force is applied to something, there is an equal and opposite reaction force which acts to push the pusher in the opposite direction (trolley stops, indignant shopper falls over).

These forces don't apply only in supermarket aisles, but are 'universal'. That is they apply everywhere, even in space. Because these forces are encountered in everyday life, they seem obvious. So it's easy to underestimate the genius behind the man who first defined them, not only in words but in mathematical form.

In a similar way, Newton developed a theory of gravity, which he also described as a force. Is there a connection between these two ideas? Newton's experience with a falling apple gave him the insight to describe, for the first time, the law of gravity which is so central to our story. The story goes that his theory of gravity came to him when he saw an apple fall from a tree in his mother's orchard. What is the connection, Newton asked himself, between that fallen apple and the Moon circling the Earth? Why doesn't the Moon fall to the ground like the apple? Newton decided that the Moon *was* falling to the ground, but at the same time it was trying to travel in a straight line like a cosmic shopping trolley. The gravitation of the Earth was pulling the Moon in a perpetual curve around itself, just like you pulling a fully-laden trolley around a corner.

From the force point of view the Moon is held in place by the pull of gravity which provides exactly the correct force necessary to deflect it into a circular orbit. It is accelerating towards the Earth, like a falling apple, but that acceleration is just what is needed to create a circular path. Thus the Moon is in a constant state of freefall.

Think of it this way. If you threw an apple as hard

as you could, you would see it fly through the air. But no matter how hard you throw it, the apple always curves back down to the ground. Now the Earth is round, so imagine the curvature of the Earth also curving down in the distance. When you think about it, the only reason the horizon exists at all is because the Earth is round and you can't see 'over' the curve of the Earth. But, if you were able to throw the apple hard enough, you just might be able to throw it over the horizon. It would still land, of course, but for a short way its trajectory would match the curvature of the Earth.

Well, said Newton, what if you threw the apple just hard enough that it fell towards the surface of the Earth at exactly the same rate that the ground fell away beneath it. The apple would be constantly falling, and the ground forever falling away. The apple would then circle the Earth and could eventually return to its starting point and hit you in the back of the head. This, said Newton, is exactly what the Moon is doing. In fact, when Newton compared the force needed to keep the Moon in orbit with the force needed to accelerate the falling apple towards the ground in the orchard, he came up with almost the same figure. Gravity is universal. Following this logic, Newton discovered his law of gravitation.

Gravity, he said, acts between all objects with a force proportional to their masses and is inversely proportional to the square of the distance between them. According to Newton, everything has a gravitational attraction. You, us, this book and the Earth all have gravity. If we didn't, we'd just float away. Newton's law also says the closer two objects are together—an apple and the Earth, for example—the greater the pull of gravity and the greater motion needed to keep them from falling toward one another. For example, the Moon's motion through space—its desire to travel in a straight line—is just balanced by the mutual pull of gravity between it and the Earth.

Newton's concepts are based on forces. He showed that the motion of objects—shopping trolleys, cars, trains and planets—could be explained by the sum of various types of forces. When different types of forces act, the resultant motion is the sum of the effect of each different force. Especially in the case of gravity, we have a law which epitomises the concept of laws in physics. Newton's laws apply equally everywhere in the Universe.

It has been said that Newton's ideas on gravitation are simply an approximation. While this is true, it is important to emphasise that the approximation is an extremely good one. Certainly it has been good enough to build aeroplanes, skyscrapers and trains. It also proved good enough to explain the insights of another brilliant social misfit, a Polish cleric named Johannes Kepler, who was the first to correctly explain the nature of the orbits of the planets around the Sun. In Newton's theories we find an explanation for Kepler's observations. We also find the power to discover new planets in the solar system.

In 1781, the planet Uranus was discovered by William Herschel, an English astronomer. Uranus was the first planet discovered since antiquity and was therefore a major event in our exploration of the Universe. Yet the discovery was made completely by chance: Herschel was simply charting the heavens the way sea explorers charted the surface of the Earth. One night he just happened to spot the planet and, although he first thought it was a comet, he knew simply by its appearance that it was not an ordinary star. Herschel and others began plotting the new planet's movement against the background stars in an attempt to define its orbit around the Sun, as had been done for the other planets.

Newton's laws of gravitation worked wonderfully here. According to the best measurements of the day, each of the planets moved in strict accordance with Newton's laws. The new planet, later to be named Uranus, turned out to be as unconventional as some of the astronomers who studied it. No matter how the celestial mechanists fiddled the numbers, Uranus just wouldn't follow its predicted path among the stars.

Here was a problem. Could it be that Newton's gravitation had a limited range, and that beyond a certain distance from the Sun it broke down, allowing planets to wander unleashed throughout the starry sky? Perhaps that's an exaggeration, since the amount that the observed position of Uranus differed from its predicted position amounted to the equivalent of the width of a hair from your head seen from a distance of a hundred metres! Yet this tiny amount annoyed the astronomers of the time like nothing else. What was causing the error?

Enter two bright, young men who each had a flair for mathematics. One was British, John Couch Adams; the other was a Frenchman, Jean-Joseph Leverrier. By the early 1830s, the problem of Uranus' wanderings had become so pronounced that astronomers began to wonder whether it might be caused by the presence of another planet further yet from the Sun. Such a planet would exert a gravitational pull on Uranus, tugging it from its predicted location. Adams was the first to make an accurate estimate of the unseen planet's mass, distance from the Sun and, most importantly, location in the sky. By October 1843, Adams had a reasonable estimate of where in the sky an observer might find the new planet, but owing to petty snobbery and personality differences, he could not gain the interest of the British Astronomer Royal at the time, George Biddell Airy. Adams's predictions remained untested.

Two years later on the other side of the English Channel, Leverrier performed similar calculations to Adams, quite unaware of Adams's results. Leverrier completed his work on the 18 September 1846 and passed it on to the German astronomer Johann Gottfried Galle. Galle had, quite by chance, recently acquired a new set of star charts covering the area of sky containing the predicted position of the new planet. He began looking for the new planet and found it only a few days later, on 23 September 1846 . . . within a degree of the position predicted by Leverrier!

Controversy raged over who should be given credit for the discovery of the new planet, later to be called Neptune. James Challis, Airy's successor as Astronomer Royal, claimed he had found Neptune during his own searches but hadn't had time to verify his discovery, while it was Galle who had been the first to positively identify Neptune through a telescope. In the end history has bequeathed the credit jointly to Adams and Leverrier, although Adams received little recognition in his own time.

This was a major triumph for Newton's gravitation. It was the tiniest of discrepancies that led to the investigation in the first place. Gravitation was universal, and Newton's mathematical description seemed unstoppable. Using Newton's space, astronomers went on to discover thousands of minor planets and plot the motions of hundreds of comets.

While one of the outermost planets in the solar system confirmed Newton's gravity, the innermost planet was to highlight its limitations. Like Uranus, the planet Mercury showed slight deviations from its predicted position. This time, however, the problem was to take more than a telescopic search to solve the problem. In fact it was left to Albert Einstein to explain just what was going on.

Einstein's view of gravity was entirely different from Newton's. Where Newton thought of space as a perfectly empty void through which the planets moved in time, Einstein regarded space and time together. This four-dimensional space-time may seem rather odd at first, but it's not that hard to understand. After all, whenever you travel to work you're experiencing four dimensions: the journey through space and the time it takes to complete that journey. At any point in your trip, you can specify your position as three co-ordinates of space and one of time.

Einstein described his ideas on gravity in his general theory of relativity in 1915. In a nutshell, general relativity says that space is curved due to the effects of the matter it contains. Matter tells space how to curve, while the curvature of space tells matter how to move. The key to Einstein's theory of gravitation is the idea that mass can 'distort' space-time. Newton regarded the Moon as being constantly pulled toward the Earth, whereas Einstein said the Moon followed a natural curvature in space created by the mass of the Earth, like a marble rolling around a roulette wheel.

Let's look at the difference between the two views of space using an analogy. Newton thought of space as being like a rigid stage on which the planets rolled about in a celestial ballet, interacting with one another by their mutual gravitational attraction. Einstein, on the other hand, thought of the stage as being like a large rubber sheet stretched taut—stiff, but deformable, like a trampoline. Einstein said that the planets deformed the fabric of space. Indeed, if you place a bowling ball in the middle of such a rubber sheet it would cause the sheet to sag in the middle. This is matter telling space how to curve. Let's assume the bowling ball represents the Earth. Now, onto the sheet we roll a cricket ball to represent the Moon. The cricket ball will naturally roll towards the bowling ball because of the downward curve of the sheet. This is like curved space telling matter how to move. If you roll the cricket ball hard enough and in just the right direction it will roll around the bowling ball following the curvature of the sheet. Although it's unlikely you could do it in practice, in your mind you can picture the cricket ball endlessly orbiting the bowling ball, following the curvature of the rubber sheet but not falling in due to its momentum. Astonishingly, as Einstein proved, such orbits were actually *straight lines* in curved space-time. The Moon follows the shortest path through the curved space-time created by the mass of the Earth. The same is true for the Earth moving around the Sun, and for the Sun moving through the Galaxy.

'What's the difference?' you ask. Surely the end result is the same? Not quite. Because geometry is different in curved space, all the mathematics is subtly altered, and the predictions of Newton and Kepler are no longer quite correct. It's like the roads following the gentle curves of the hills and valleys described in the last chapter. It is space-time that is curved and not the trajectory of the Moon or the planets or, for that matter, a ray of starlight. All freely falling objects follow a straight line in space-time.

Let's look at this in a little more detail. A straight line is defined as the shortest distance between two points. For example, if you marked two dots on a piece of paper laid out flat, you could draw a line using a ruler joining them. But if you took that piece of paper and bent it the line would no longer be straight. Or would it? If you are confined to the surface of the paper, as we are to space, it still represents the shortest distance between the two dots, so by definition it is a straight line.

To put it in a more realistic setting, think about the curve of the Earth's surface on which the shortest distance between two points is a great circle. Let's say you board an aircraft flying from London to Sydney. During the journey the plane does not follow a straight line in the usual sense, but travels along a great arc parallel with the surface of the Earth. To pass the time, you pull out the free copy of the in-flight magazine. Turning to the back page, you see a map of the Earth with the various destinations offered by the airline. You can see that, when the map is projected onto a flat piece of paper the routes taken by the aircraft are anything but straight. The curvature is an illusion created by our dependence on flat paper. And yet the plane is flying the shortest possible route between London and Sydney.

Why is it so surprising—even uncomfortable—to think of space as being curved rather than flat? It's simply prejudice: we're not used to the idea of space being curved by massive objects. Newton in all his genius thought space was flat and that all objects in it were simply attracted to each other by the force of gravity. We call this 'common sense' science because it's easy to relate to everyday experience, like wrestling with a shopping trolley. But Einstein said that it's more accurate to think of space as being curved and bent by the objects which inhabit it. Einstein's theory of gravity says that matter tells space-time how to curve. Just as the bowling ball creates a depression in the rubber sheet, the Earth creates a dimple of curved space-time around itself, around which the Moon travels the shortest possible path.

The most important thing to keep in mind about Einstein's Universe is the fantastic stiffness of space—of the rubber sheet, if you like. Let's put it into perspective: let's say the magnitude of the stiffness of a rubber sheet is about 1. Using this criterion, the stiffness of solid steel is about 100 000 000 000, or 10^{11}. Space has a magnitude of about 10^{43}, a one with 43 zeros after it! Space is a billion billion billion times stiffer than steel! In Newton's space, you just have to set this number to infinity—this makes space infinitely rigid and the curvature non-existent. In other words, the enormous but not infinite stiffness of Einstein's space-time tells us that, while space is not infinitely rigid, it is very, very, very rigid. In fact, odd as it sounds, space is the most rigid stuff in the Universe.

Surely all this is just playing with words? Describing

planets as rolling around the local curvature in space is just an alternative way of explaining the same thing Newton did by saying gravity was an attractive force that deviated planets from moving in straight lines. Not quite, and this is where Mercury comes in.

While Newton's gravitation failed to explain errors in the predicted position of Mercury, Einstein's gravitation succeeded. In 1846, the same year Galle spotted Neptune for the first time, Leverrier noticed that Mercury also behaved in an odd fashion. Planets move in oval-shaped orbits, passing closest to the Sun at one point in their orbits and farthest from it at another. The point of closest approach to the Sun is called the perihelion. Leverrier noticed that the position of Mercury's perihelion was moving in an unexplained way. The unexplained movement was incredibly small: out of its 360-degree orbit around the Sun, the movement amounted to 43 seconds of arc per century. This is a movement of about one-sixth of the apparent size of the full stop at the end of this sentence held at a comfortable reading distance, every hundred years. Nonetheless, each of the other planets now had its orbit defined to such accuracy, and Mercury was refusing to fall into line. Aha! thought Leverrier, here's another undiscovered planet. He was so sure of the existence of this planet which seared in the heat of the Sun that he named it Vulcan, after the Roman god of fire.

Leverrier never did find Vulcan, for the simple reason it didn't exist. In fact it was seventy years before the problem of Mercury's orbit was solved. Because of its proximity to the Sun, Mercury is deep within the gravitational well created by our local star. While Newtonian gravity is sufficient to explain the motions of the other planets to a high degree of accuracy, little Mercury's orbital irregularities refused to be solved by Newton's flat space. Using the idea of curved space, however, Einstein explained the observed shift in

Mercury's perihelion. Vulcan evaporated; general relativity triumphed.

This wasn't the most spectacular demonstration of general relativity, however. Einstein was not yet famous. That had to wait for the confirmation of another strange prediction: not only does matter tell space how to curve, it also distorts the passage of light from distant sources such as stars. Before we look at how general relativity explained Mercury's odd behaviour and how gravitation can alter the passage of starlight, we need to look at Einstein's theory in a little more detail.

Chapter 3
A Theory of some Gravity

General relativity is based on the concept of space-time, a concept so fundamental to understanding gravitational waves that we need to spend a little more time exploring Einstein's strange Universe. Einstein derived his general theory of relativity with very little experimental data to support him. He began with two basic ideas. The first was that gravity and acceleration are indistinguishable. The second was that matter in freefall always takes the shortest possible path in curved space-time.

These are subtle but powerful ideas. The first, that gravity and acceleration are indistinguishable, is called the Equivalence Principle. Earlier we saw that Newton defined gravity as an attraction between any objects which have mass, that is, everything. But have you ever wondered what, exactly, is 'mass'? Mass can be measured in a couple of ways. One is familiar to almost all of us: using a set of bathroom scales. Scales like these measure the amount your mass is pulled downward by the Earth's gravity against a stiff spring. The degree that the scales are depressed (indicated by the dial) expresses the gravitational attraction between you and the Earth. This is your gravitational mass. Now Newton's law of gravity says that gravity is very simple and depends only on the masses of the objects involved, in this case, you and the Earth. Gravitational mass is also independent of composition: it doesn't matter whether the masses are electrically active or neutral, hot or cold, solid or liquid, alive or inert, muscle or fat. No matter what you feel or wish about your weight, your scales will measure the same gravitational mass. The only simple way to alter your gravitational mass is to eat more or less, depending on the desired outcome.

There is another way, however. Move to another planet! Gravitation is a measure of the attraction between two objects, and this depends on their masses. So, if you want to be lighter without dieting, you can always try moving to a planet with less mass and hence less gravitational attraction. Mars, for example, or better yet the Moon, both of which are much smaller than the Earth and so have less gravitational pull. However, you'd probably find that the living conditions there aren't as pleasant as they are on Earth, so you might be better off reducing your own mass, rather than that of your home planet.

There's a second way of measuring mass, however, and it has to do with motion. More specifically, with acceleration. It is the mass which makes it difficult to accelerate and steer a heavily laden shopping trolley: it is called inertial mass. Perhaps the most useful analogy to explain the relationship between gravitational mass and inertial mass is a lift or a spaceship. You must have felt the strange sensation of that first second or two when a lift starts moving upwards and you feel slightly heavier for an instant. Even better is the sensation of lightness when the elevator begins its controlled descent. What you're feeling is a change in your 'weight' due to acceleration of your inertial mass. But to explain the difference between gravitational mass and inertial mass, let's board an imaginary spaceship.

Think about being in a spaceship and imagine that you brought along your bathroom scales. When the spaceship is floating idly in space it's in a state of freefall, and if you were to stand on the scales they would read flatteringly little. In fact, you would 'weigh' virtually nothing. In freefall all you have to do is flex your toes against the floor and you'll float to the roof of the cabin. Because you and the spaceship are falling at the same rate, neither is subject to any external force. When you flex your toes, you are demonstrating one of Newton's laws of motion: you stay still until you move a muscle and exert a force.

Now, while you're standing on the scales, reach over and press the spaceship's start button (this is a very simple spaceship). As the rockets fire and the spaceship starts to move, you can see the reading on the scales begin to climb. Again, Newton's laws in practice: your body wants to stay still while the rocket begins to move, so you exert a force down on the scales. This is exactly the same force you feel in your car when you accelerate. As the spaceship accelerates you can see your 'weight' increase. What the scales are measuring now is your 'inertial mass'. It's called inertial mass because it is a measure of your mass based on your inertia, your tendency to stay still. Every time you accelerate in your car you feel your inertial mass pressing you back into the seat. In the spaceship, you feel your inertial mass on the soles of your feet as the spaceship accelerates.

As the rate of acceleration increases, the reading on the scales also increases. If you glance down at the scales when the spaceship reaches an acceleration of one g, or 9.8 metres per second per second, your 'weight' will be exactly the same as it was back on Earth. If the spaceship accelerates at a faster rate, your 'weight' will increase. When the spaceship is accelerating at 9.8 metres per second per second, you feel the equivalent of one 'Earth gravity', or one g. Hang on to this thought for a moment.[1]

At 19.6 metres per second per second—that is, 2g—you feel twice as heavy as you do now, as if you were carrying another person while standing on the scales back on Earth. At this point, you reach over and press the stop button and the acceleration begins to decrease as the rockets die down, even though you are still hurtling through space. Before long, you will be lighter, back to your weightless self. Notice how similar this is to accelerating in a car. Once you stop accelerating, even though you might be travelling at tremendous speed, you feel normal again.

If the acceleration stayed at 9.8 metres per second per second or thereabouts, you could freely walk around your spaceship and, if it had no windows, you could imagine yourself back on Earth. In fact, there would be no way of telling that there was anything different—the spaceship might as well be in a parking lot as in deep space. That acceleration of 9.8 metres per second per second is the equivalent of one g, that is, one Earth gravity. This is an important point: you cannot tell any difference between your weight on Earth and the weight you feel in the accelerating spaceship. It is impossible to distinguish any difference at all. This is one of the most important principles in science: inertial mass and gravitational mass are indistinguishable.

The identity between inertial mass and gravitational mass is known as the equivalence principle. But why should inertial mass and gravitational mass be the same? Galileo was the first to be surprised by this, and ever since, physicists have tested the equality of inertial mass and gravitational mass to greater and greater precision. Early in the twentieth century the Hungarian physicist Baron Lorand von Eotvos set up an elegant experiment to check whether inertial and gravitational mass were really equal. He found that they were identical to ten parts per billion. In the 1970s, Robert Dicke of Princeton University and Vladimir Braginsky of Moscow State University checked the equality to within one part per hundred billion and one part per thousand billion, respectively. It won't be long before space-based experiments will check it to even greater accuracy. The reason why physicists want to test the equivalence principle to even higher accuracies is that some theories predict there will be a difference . . . perhaps at a level of one part per million billion, or even less.

There was a bit of a scare in the 1980s, however. A scientist at the University of Queensland published results implying that all gravity experiments, including those used to test the equivalence principle, were wrong. All previous experiments had been corrupted by a new force in nature, called the 'fifth force'.[2] Other work purported to show that Baron Eotvos's experiments had also been affected by a force which acted differently on the masses used, depending on their composition. These suggestions struck at the very foundations of general relativity. Not surprisingly, they generated an enormous resurgence of classical gravitational experiments—elaborate improvements of the original tests.

Eventually, however, the results indicating a fifth force were found to be in error, but in a way that illustrates just how delicate these experiments are. The experiments that 'demonstrated' a fifth force relied on an accurate survey of the strength of gravity over a Queensland mine site. The problem arose because the gravity surveyors liked to stay on roads, avoiding hollows and other inconvenient spots. This subtle bias was enough to completely alter the interpretation of the results.

Today, all signs of a fifth force have faded away. The equivalence principle is unchallenged except for some predictions of infinitesimal violations. For all practical purposes, inertial mass and gravitational mass are identical, and gravity makes no distinction between the nature and composition of the masses.

Now think about freefall. Let's take a cricket ball and a bowling ball to the top of an office building. We've cleared the street below with the help of the police, so it's safe to drop them at the same time. Now, just before we drop them, think about their masses. The cricket ball has less mass, and so it's easier for gravity to start it moving when we let it go, just like the empty shopping trolley in Chapter 2. The bowling ball, on the other hand, has much more mass and so is harder to start moving, like the full shopping trolley. At the same time, the gravitational mass—the weight, if you like—of the two balls is quite different. Intuition might tell you that the bowling ball will fall faster because it's heavier than the cricket ball. Let's see.

When the balls are dropped they fall away, watched carefully by a crowd of curious onlookers. Within seconds they hit the ground: one bounces, the other shatters into numerous pieces. But the important thing is they hit the ground at the same time![3] What happened? The gravitational mass of the bowling ball is greater than that of the cricket ball, you're quite right. But its inertial mass was also greater—in fact identical to its gravitational mass—so it resisted acceleration due to gravity. The cricket ball, on the other hand, had less gravitational mass and so tended to be pulled less by the gravity of the Earth. At the same time it had less inertial mass to resist the pull of gravity. The end result is that gravitational mass and inertial mass tend to cancel out. This means the rate of freefall and the trajectories of moons and planets are independent of mass. Galileo is said to have carried out this experiment using weights dropped from the Leaning Tower of Pisa almost four hundred years ago, though it is probably just a story. He did, however, carry out similar experiments using weights rolled down gently sloping planes.

The second key concept in general relativity is space-time. According to Newton, natural trajectories are straight lines at a constant speed, and gravity is an external force which creates deviations from this natural trajectory. The acceleration of a falling apple is the result of an external force—gravity—acting upon the apple. In Einstein's curved space-time, however, the falling apple is the natural state of things, and its prior attachment to the tree is the force needed to prevent it from following the curvature of space-time.

Looked at another way, Newton was saying that space is like a flat sheet of wood. If the Earth were represented by a nail hammered into the wood and the Moon represented by a marble, gravity is like a piece of string joining the nail and the marble, preventing the marble from rolling off into space. Einstein said that there is no string nor a flat surface. According to him, the space surrounding Earth is more like a salad bowl, and the Moon slowly rolls around near the rim following a contour. Putting it yet another way, gravity is not the force we feel when the elevator cable breaks, but the force we feel when the elevator is stationary, and the cables are acting to prevent our own natural freefall.

As we've seen, the evidence Einstein had for these claims was slim: a tiny deviation in Mercury's orbit. Before we can fully appreciate how even this prediction was evidence in Einstein's favour, we have to think a little more about what it means for space-time to be curved. Specifically, we must find out how to measure space-time.

The fact that we talk about space-time and not just space should not be any bother. Don't try to imagine four-dimensional cubes, whatever you do. Just as architects usually reduce house plans to a set of two-dimensional space-space diagrams—plan, side elevation and end elevation—so we can reduce space-time to a set of two-dimensional space-time diagrams: height versus time, length versus time, depth versus time, and so on. Unfortunately we don't have the luxury of being able to make a simulated picture of space-time equivalent to the architects' perspective drawings, but a movie is rather similar. A movie is a time-sliced set of space-space diagrams which create a space-time simulation! Space-time is no more complex than that, and curved space-time is like projecting a movie onto a curved screen. The curved screen creates a distortion of the geometry of space. Conversely, when you see the distorted pictures on the screen it tells you that the screen—that is, space—is curved. Although the curvature of space-time is beyond human experience, we can at least see examples of the geometry of curved space all around us: the distorted reflections in polished saucepans and car hubcaps, ripples in old window panes, and the clever page-turning simulations used on television.

In this world of curved space, the geometry that applied to flat space no longer works. Think about a circle drawn on a piece of paper. It has a certain area that can be calculated. Now think of a circle with the same diameter drawn on a ball. Because the surface of the ball is curved, the surface area inside the circle is larger than it is for the same-diameter circle drawn on a flat piece of paper. The increase in area depends on how curved the surface is: the greater the curvature, the greater the increase in surface area compared with a flat surface. However, in general, circles drawn on spherically curved space always have greater area compared with flat space.

This extra area tells us that space is positively curved. Notice that the radius of the circle on the curved surface must be measured on the curved surface itself. In our two-dimensional examples we may be tempted to cheat because we look at it from a higher dimension—from above the ball—but in true, four-dimensional space-time we can never step up into another dimension to get the sort of 'God's-eye view' we have when we do geometry on a ball.

A second way of measuring curvature is to measure the sum of the angles in a triangle. In high-school geometry we're told that the sum of the angles in a

Figure 3.1 A triangle drawn on a spherical surface can have three right angles. Such a triangle has a total of 270 degrees.

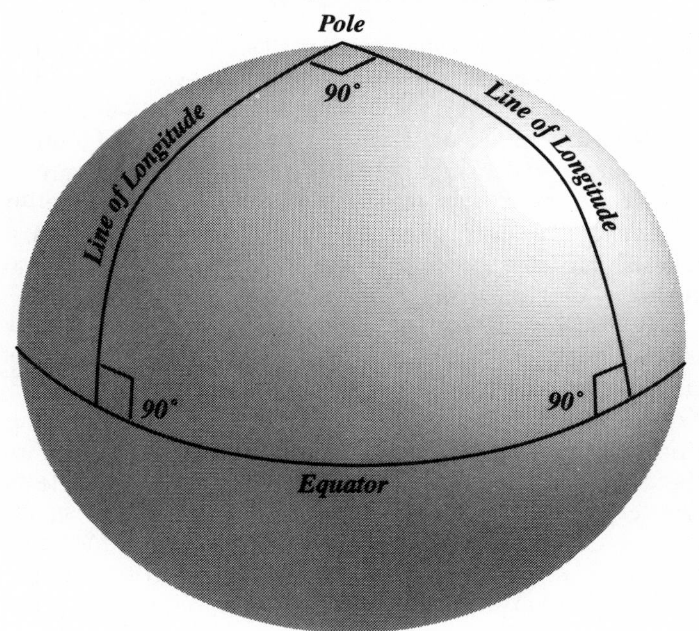

triangle equals 180 degrees. Now, think about a triangle drawn on a globe of the Earth. Start on the equator and follow a line of longitude until you reach the north pole. From there follow a line of longitude 90 degrees to the first one until you reach the equator again. Now turn 90 degrees and follow the equator around to where you started from. What you've just traced is a triangle with three right angles, a total of 270 degrees! Clearly, the geometry of curved space violates the rules of flat space. The excess degrees again tell us that we're dealing with positive curvature.

But there are other curves we can follow. Think about a horse saddle with the curve going up in one direction and down in another. If you draw a triangle on this type of surface you'll find that the angles add up to less than 180 degrees. This is negative curvature. But here again the usual rules of flat space don't apply. You should keep in mind that the differences we've been talking about are very small when the triangles and circles we draw are small, or if the curvature is slight. Only on much larger scales, or where the curvature is pronounced, do the differences become important. Incidentally, this is why Newton's physics is still important and why it reigned supreme and unchallenged for over two hundred years.

By now you may be saying something like 'Surely this geometry is only true of curved two-dimensional surfaces, not the four-dimensional space-time we inhabit?' Well, you're half-right. We are only measuring two-dimensional surfaces which we can examine in three-dimensional space. We can pick up the ball and saddle, examine them and compare them with flat space. But we can do this because we have an extra dimension. If you were a two-dimensional creature living on the surface of that ball or saddle, you'd be incapable of doing the sorts of comparisons we've just made. All you would know is that in your curved, two-dimensional universe, the angles of a triangle don't add up to 180 degrees.

OK, now that you understand the methods for checking for the existence of curvature, we can step back into four-dimensional space-time and test for curvature here. Remember how we said that these effects are obvious only over large distances? Well, we need some way of checking the angles of a triangle drawn on a pretty big scale—say, around the Earth. To make life simple and ensure that we are indeed drawing straight lines, we'll use lasers mounted on three satellites placed at roughly equally-spaced points around the Earth. Further, each of these satellites is visible to the other two.

From each satellite we point a (non-destructive!) laser at the other two. From each satellite, we carefully measure the angle between the apparent position of the incoming laser beam. Lo and behold, when we add up the angles measured at each satellite, we get 180.000 0001 degrees!

Still sceptical? Still arguing that it isn't the curvature of space at all, just the bending of light rays by the gravity of the Earth? We won't disagree with you. Three curved laser beams will trace out a triangle whose angles add to more than 180 degrees. But think about it for a minute. When we were measuring the angles on the ball and the saddle, we were using models that we could examine from the luxury of our God's-eye view of a third dimension. We can't do that in space. We can't step into a higher dimension and compare the space we live in.

Nor do you have anything more rigid to confirm the curvature of the laser beams. No matter what you try, nothing would resist the pull of gravity—even a steel railway line would sag and twist in space. Light traces out the ultimate straight line. Nor can you make little triangles to measure the curvature of space along each straight line because the curvature becomes vanishingly small as the triangles shrink. Curvature is not a local property, it is a global property. You cannot step outside the Universe and examine it. Curvature exists everywhere.

You can, if you want, continue to think of this curvature of space-time as simply the deflection of light beams under the effects of the Earth's gravity. However, it is much more convenient to do away with gravity and accept that the laser followed the curvature of space-time. More importantly, as we'll soon see, the curvature of space-time predicts many subtle phenomena, which have now mostly been tested. These phenomena cannot easily be explained by a simple force, but are beautifully explained by curved space-time.

You may have noticed that so far we have said little about time, other than to mention that it forms a fourth dimension to the usual three dimensions of space. While we won't attempt to explain in detail Einstein's conclusions about time, we will at least present them for you to ponder. Space and time are inseparable. An object not falling along its natural path through space-time is ageing more slowly than one that is. Bizarre as it sounds, a stone sitting on a cliff is ageing more slowly than if it were to fall off the cliff and follow its natural path in space-time. Einstein said that freefall trajectories are the shortest possible trajectories in space-time. We saw evidence of this in the curved paths of laser beams fired between the satellites we put in Earth orbit. Clearly, if the stone sits on the cliff, that is its shortest possible route in space alone. But in space-time, things are different. Everything is always going somewhere in time—after all, everyone of us is ageing. But because the stone's natural path through space-time is being obstructed by the cliff, it's taking longer to get to its destination—old age—than it would in freefall. To put it succinctly, the stone sitting on the cliff is ageing more slowly than if it were falling towards the crashing waves below.

Furthermore, think about what causes space-time to curve. It is gravity, the gravity of objects like the Earth and the Sun. So if space-time curvature alters time, and curvature is caused by gravity, then gravity alters the passage of time. According to Einstein's view, when an object is in freefall everything is simple. Time is normal. But when something interrupts the natural freefall of an object—such as the cliff holding up the rock—time goes more slowly.

The change in the passage of time is called gravitational time dilation. It tells us that if you raise a clock one metre above the ground, its rate will be altered by about one part in 10^{16}. If you're reading this on top of a mountain and you've lived there all your life, bad luck! Your life—perhaps two to three billion seconds of it—will be about 100 microseconds shorter than if you moved to a coastal town. In 1916 Einstein never imagined that real clocks would ever be accurate enough to measure this effect. However, modern atomic and sapphire clocks are so accurate that gravitational time dilation starts to be an annoying problem!

Now we can turn things around and say: if gravity distorts time intervals by the amount that general relativity predicts, this is good evidence that space-time is curved by matter.

So far, we have made some pretty outrageous claims about the state of the Universe. Can we substantiate any of it? Well, yes, quite a bit. We won't attempt to draw you into a mathematical discussion to prove the curvature of space-time and time dilation, but we can present you with modern observational evidence that what we've been telling you is the truth. We've already mentioned one example of confirmation of general relativity, Mercury's perihelion, and we've hinted that direct measurements of time dilation have been carried out. But there are bigger, better examples to talk about.

It's now time to step out into Einstein's Universe and start looking at some of the amazing phenomena that can only be described by Einstein's general relativity. We won't be doing away with Newton and his gravity, however. The reason is that, as we sail this cosmic sea in search of gravitational waves, it is easier to explain the nature of some of the bizarre sights we'll see using metaphors from both Newton's and Einstein's theories. We'll talk about the deflection of light rays by gravity alongside comments about light rays following the curvature of space-time; in one breath we'll talk about the force of gravity acting on objects and in the next describe how those objects create curvatures in space-time. The reality is that it's easier to work with mixed metaphors.

The progress from Newtonian physics to relativistic physics is one of successive approximations, and the crudest one is often sufficient. Perhaps some of the mixed metaphors represent our difficulty in fully assimilating relativistic concepts. If so, let's hope this book and the detection of gravitational waves help imbed the true nature of space-time into our education system.

Notes

1 Acceleration is defined as the rate of change of speed. Your car's speed is measured in kilometres per hour. If you were travelling at 60 kilometres per hour, your car would travel a total distance of 60 kilometres after one hour. Acceleration is a measure of how your car's speed changes with time. Say you were sitting in your car about to drive off. As you press your foot to the accelerator your car speeds up, that is, accelerates. You can feel the change in speed as you're pressed back into the seat. If you accelerate at ten kilometres per hour per second, then your speed increases by ten kilometres per hour at every tick of the clock. At the end of the first second, you would be travelling at ten kilometres per hour. By the end of the next second, you'd be travelling twenty kilometres an hour, thirty kilometres an hour at the end of the third second, and so on. At the end of eleven seconds you've reached 110 kilometres an hour, the national speed limit—stop accelerating! This acceleration would be described as 10 kilometres per hour per second. In other words, your car's speed is changing by ten kilometres an hour every second. The same method of describing acceleration works with other units of measurement, such as metres per second per second.

2 The four fundamental forces of nature are gravitation, electromagnetism, and the strong and weak

nuclear forces. These are thought to be the basis of all matter and energy in the Universe.

3 We assumed no wind resistance! This experiment works properly only in a vacuum.

Chapter 4
The Cosmic Looking-Glass

General relativity is a spectacularly successful theory. One of the reasons it is so famous is that it has survived half a century of rigid tests. Alternative theories have been proposed over the years, but Einstein's theory reigns supreme. So far we have explored the basics of general relativity and have a working knowledge of some of the principles involved. But how do we know that it is true? How do you confirm such a bizarre view of nature?

As we've seen, one of the major successes of general relativity was the explanation of the advancement of Mercury's perihelion. But exactly how does general relativity do this? In the early seventeenth century, the great Danish astronomer Johannes Kepler was the first to describe planetary orbits correctly. Until then, planets were thought to orbit the Sun in perfect circles, a left-over from the days when planets were thought to be divine. Since a circle is considered to be a pure shape, it seemed only fitting that planets moved in circular orbits. Kepler gave us a number of laws of planetary motion, one of which was that planets move around the Sun in ellipses, with the Sun at one focus of the ellipse. Another was that, as a planet moves through its orbit, it sweeps out equal areas of the ellipse during equal time periods. A planet moves faster at its closest point to the Sun and slower when farthest from the Sun. Because of this speeding-up and slowing-down, each planet sweeps

Figure 4.1 In the left circle, the area is defined by the perimeter. In the right circle, drawn in curved space, there is always a little extra surface area left over after the circle has been completed.

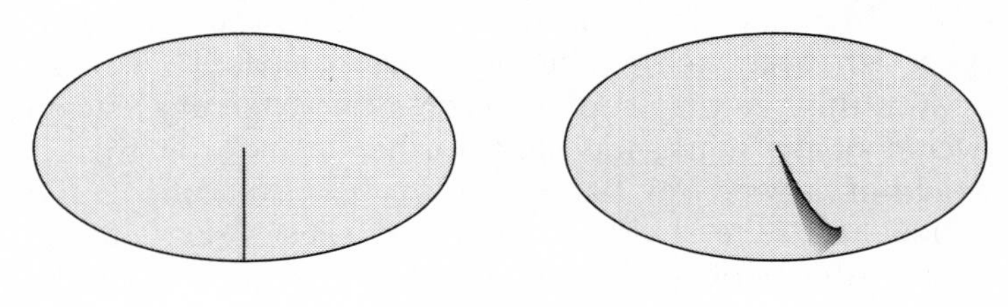

out an equal area of its orbit during each day, week or month, no matter where it is in its orbit.

Thanks to Einstein, we know that each planet's orbit is through curved space, space that is curved by the Sun. You may remember that one way to measure this curvature is to measure the perimeter of a circle and compare it with its surface area: excess area is a sign of positive curvature. The same goes for an ellipse or any other shape. By comparing how the area of an ellipse in curved space differs from one in flat space, we can measure how curved the space is. You can illustrate this point by drawing an ellipse on a soccer ball and an ellipse of the same size on a flat piece of paper. The ellipse drawn on the soccer ball has greater area because of the curvature of the ball's surface.

The same difference in area applies to the space marked out by the orbits of the planets. Because the space that Mercury moves through so near the Sun is so steeply curved, the ratio of the perimeter of its orbit and the area traced out by the orbit is noticeably different from the same ratio in Newtonian flat space. When you measure the area traced out by the planet based on the distance it has travelled around the perimeter of the orbit and compare it with the area based on the curved space, they don't match. The area in curved space is always a little more.

In reality, what happens is that, at the end of every orbit, Mercury has a little bit of its journey through space left over, due to the curvature of space. To make up for this, at the 'end' of each orbit Mercury has to overshoot the mark, travelling further in its orbit than it would if it were in flat space. As a result, the entire ellipse tends to twist a little on each orbit of the planet—the perihelion point where the planet is closest to the Sun moves around just a little each time—so that after many many cycles, the orbit traces out a pattern that looks like the petals of a daisy.

This movement of the point of perihelion is called precession, and all of the planets do it. When the curvature is slight, as for the planets far from the Sun, the difference is minuscule. By the time you take into account all of the various gravitational tugs and pulls exerted by the other planets, the perihelion precession of the outer planets is completely swamped by these larger effects. For tiny Mercury, deep within the gravitational well created by the Sun, the difference is significant. Mercury's perihelion precesses by 43 seconds of arc—one-fortieth of the diameter of the full Moon—every one hundred years. At this rate, it will take Mercury's perihelion point four million years to complete one rotation of the Sun. The Earth, incidentally, does the same thing every thirty million years. Small as these figures might be, they are enough to confirm Einstein's general relativity.

That wasn't the most spectacular demonstration of general relativity, however. One of the many bizarre predictions of the theory is that light beams can be bent or deflected by gravity. The idea goes back to the similarity of acceleration and gravitation. Let's say we're in our spaceship again, accelerating through space. If you stand on one side of the cabin and turn on a torch, what happens? The photons that make up the beam set off on their way across the cabin at the height of the torch. Even though the speed of light is immense—300 000 kilometres a second—it's not infinite. Because the spaceship is accelerating, by the time the photons make it over to the other side of the cabin, the spaceship has moved on a little and the photons strike the opposite wall of the spaceship a little below the height of the torch. Looked at side on, the beam from the torch appears slightly curved. As you can imagine, owing to the speed of light, the amount of deviation at the small accelerations that we're used to is minuscule. However, given enough acceleration, the effect should be measurable.

Now, Einstein reasoned that, since there is no difference between acceleration and gravity, gravity should also distort a ray of light. In other words, gravity affects more than just the movement of stars and galaxies; gravity also distorts their light. But in the same way that you need an immense acceleration to demonstrate this bending of light, the mass of an object capable of creating enough gravity would have to be immense. While all objects create their own curvature in space-time, usually this curvature is tiny; so are the effects it produces, as we've seen by the success of Newton's gravity with all the planets except Mercury. How could the predicted bending of light be tested?

When the British astronomer Arthur Eddington heard of general relativity, he immediately recognised its importance and started making plans to test its predictions. To test Einstein's prediction that light could be bent, Eddington decided to use the most massive object at hand—the Sun. According to Einstein, the light from the stars near the Sun would be bent, shifting their apparent position like a celestial mirage. Because of sunlight reflecting off our atmosphere, the stars are normally invisible in the daytime sky. But during a total solar eclipse, the Moon covers the Sun, and the stars become visible in the darkened sky. This allows astronomers to photograph the stars near the Sun, and then to compare these 'apparent' positions with the stars' known fixed positions. General relativity predicted that, in the case of the Sun, a star's image would be deviated by almost 2 seconds of arc near the edge of the Sun.

Eddington organised two expeditions to observe the total solar eclipse of 1919, himself leading one of the teams to the island of Principe, off the west coast of Africa. The other expedition was sent to northern Brazil. Two expeditions were mounted to improve the chances of seeing the eclipse: should one team experience cloud during the eclipse, it was hoped that the other would have clear skies. As it turned out, both teams had clear weather. When the eclipse photographs from the two teams were measured, the displacement of the star images near the Sun agreed with Einstein's prediction.

Eddington reported his results to the public in November 1919. They were presented as a stunning confirmation of Einstein's theory. Headlines stated 'Einstein's Theory Triumphs' and 'Men of Science . . . Agog over Eclipse Observations'. The stars were deflected; Einstein was famous. Yet looking at the results of those observations today, they appear to be rather marginal. The sizes of the deflections were about equal to the errors in measurement on the photographic film. Since then, however, the observations have been repeated to greater precision, and Einstein's fame was indeed justified.

The idea is now familiar: space-time is curved by matter; light follows the curvature of space-time. The Sun creates a big dimple in space-time, and so any light beam travelling through space is deviated by the curvature of space surrounding the Sun. In a way, it looks like the distortion created when you look through a glass of water, or a magnifying glass: everything seen through the glass is distorted, only in the case of the Sun the distortion is very slight.

It took a long time to test a third prediction of general relativity: time dilation. This is the effect that gravity has on time, as we discussed in the last chapter. In freefall, a clock is perfectly well-behaved. But if you influence it in any way, either by putting it on a shelf (its freefall is now interrupted) or accelerating it in a spaceship (remember gravitation and acceleration are indistinguishable), it will run slower. Of course, if you're standing next to the clock you'll have no way of knowing, simply because you'll be ageing the same amount. We've seen that if you lived on a high mountain your life would be about 100 microseconds shorter than if you lived by the sea—gravity is a bit weaker there because you are further from the centre of the Earth. You don't need to worry, though: your thoughts, your heartbeat, your breathing, everything about you would be correspondingly faster as well. And the healthy air would probably compensate a millionfold compared with the gravity effect.

In the mid-1960s, an experiment was conducted at Harvard University to test the time dilation predicted by general relativity. For their clocks, the researchers used gamma rays. Gamma rays, or any other type of electromagnetic waves, including light, can be thought of as streams of photons, tiny 'packets' of light. The researchers used radioactive iron as a source of gamma-ray photons of very precise frequency. The frequency of the photons was measured as they were fired 20 metres up a tower. Because the photons began their journey with a specific, measurable frequency, each acted like a tiny clock. When the frequency of the photons was measured at the top of the tower, each one had a longer frequency, was 'ticking' slower, than when it began. The slower ageing of the photons agreed to the predictions of general relativity to 1-per-cent accuracy.

You can think of a photon as being just like a clock: it 'ticks' with a certain frequency. Further, the energy of a photon is governed purely by its frequency—high-frequency photons have more energy than low-frequency photons. Light photons have more energy than radio photons because they 'tick' a billion times faster; this also means that their wavelength is a billion times shorter than radio waves. If the frequency of the photons is 1000 times higher still, their wavelength shrinks to the size of atoms, and they can then pass through your body to make pictures of your insides; these photons are, of course, X-rays. When a photon loses energy, for example by climbing up a tower, its wavelength increases. As a by-product of this, its colour becomes 'redder', since red light has a longer wavelength than blue light. When a clock changes its speed with altitude, we call the phenomenon gravitational time dilation. When a photon loses energy, it's known as gravitational redshift, so called because of the 'reddening' of the photon. These are really identical phenomena. It's easy to understand how a photon could become tired climbing up a tower—clocks behave in exactly the same way.

Let's look at this in more detail, from the vantage point of our spaceship parked in orbit high above the Earth. When we look down on the Earth from space, we see it in photons emitted from the surface. In fact, everything you see around you is visible because it is either reflecting or emitting photons, that is, light. Each of these photons from Earth has had to climb out of the gravitational well created by our planet, and so has lost energy, has a longer wavelength and therefore looks redder. This is another way of saying the photons have a lower frequency and so all time intervals—time signals from radio and television broadcasts, the speed of your heartbeat and the speed of your car—will appear to be running more slowly.

Everything we see on Earth will suffer this time dilation. All time will be stretched out. Seen from space, everything down on Earth, even the rotation of the Earth itself, will be slower.

Now, suppose we use upcoming radio signals from a clock on Earth with an identical clock we have on board. The people on the ground will say that our clock is running fast, yet if they put it in a rocket and fire it up to us, by the time theirs makes it into orbit it is running at the same rate as our ship's clock. It has speeded up! But if we had synchronised both clocks before we left Earth, the one sent up later would be slow. Once time dilation has slowed a clock, it can never catch up with the one in freefall. The size of these effects is very small: the time difference after a year between a clock far from the Earth and one on the Earth's surface is about 10 milliseconds. Fortunately, modern clocks are quite capable of measuring differences much smaller than this.

There have been even more stringent tests performed on time dilation since the Harvard experiment. In 1976, Robert Vessot placed a super-accurate atomic clock called a hydrogen maser in a rocket which was flown to an altitude of 10 000 kilometres. The speed of the clock in the rocket was compared by telemetry with one on the ground. Great care was taken to measure the rocket velocity to correct for various effects, allowing the parts-per-billion frequency changes due to altitude to be measured to an accuracy of about one part in 10 000. The results agreed with general relativity to within the 0.01-per-cent accuracy of the experiment.

Today there is a set of such clocks in orbit around the Earth. They constitute the Global Positioning System (GPS), initially designed to steer missiles to their targets but now used much more constructively for world navigation. A small radio receiver on a ship picks up time signals broadcast from two or more satellites at any time. From the relative pulse arrival times, a computer calculates where you are as well as the precise time of day. The system is capable of extraordinary accuracy. Because the system relies on timing, and because light travels so fast, the timing accuracy needed to determine your position to within a metre is about three nanoseconds, or three-billionths of a second.

Clearly, you have to be pretty careful about time dilation when you need this sort of accuracy. This greatly pleased the community of physicists who investigate all aspects of relativity and so call themselves relativists. Phenomena which, while fascinating and peculiar, had always seemed irrelevant to everyday life suddenly had a practical use. Now aerospace engineers needed to know about relativity. According to an unreliable story, an engineer designing the system recognised the need for relativity, but when he did his sums he made a sign error. Instead of setting his clocks so that they ran slow, he set them to run fast. Left uncorrected, the change in gravity when the clocks were placed in orbit would have caused them to run twice as fast as if relativity had been ignored. At the last minute, a team of real relativists was brought in and just before the first launch all the clocks were adjusted.

The accuracy of the GPS is so good, in fact, that it can help guide not only US missiles to their targets, but anyone else's as well! To overcome this problem, the signals were coded in a very clever way, so that if you don't know how to decode them the best accuracy you can achieve is a hundred metres or so. During the Gulf War, however, soldiers in the desert needed to know where they were. Since there weren't enough encoded receivers to go around, but plenty of civilian ones, the US military decided to switch off the satellite coding. Suddenly there was a bonus for millions of ship-owners, prospectors, light aircraft pilots, rally car drivers and explorers, all of whom could determine their position to within a few metres. All of a sudden, one little radio could locate your position so accurately that it could tell whether you were in the kitchen or the lounge room! Imagine the outcry when the coding was switched back on after the war. Perhaps by the time you read this the coding will have been permanently switched off.

There is yet another aspect of general relativity that has yet to be tested, one of its most subtle yet fundamental aspects. When a planet or star rotates in space, it drags space-time around with it. It's a bit like a ball rotating in a bowl of water: friction at the surface of the ball tends to drag the water in the same direction as the ball's rotation, creating a swirling, spiral-shaped current. While the water near the surface of the ball rotates fastest, the dragging effect on the water dies away farther from the ball. Although invisible to all but the most delicate experiments, a similar phenomenon is occurring around the Earth right now. As it spins in space, the curved space-time it creates simply by being here is dragged around with it. This is called frame dragging. The effect is incredibly small, however, and like gravitational waves, has not yet been detected.

If the effect is so small, two questions arise. Firstly, how can you detect it and, secondly, why bother? The answer to the first question can be found at Stanford University in the United States. For more than twenty years, a group of scientists led by Francis Everitt has been developing a most wonderful and beautiful gyroscope to measure frame dragging. A gyroscope is basically a spinning top mounted on gimbals so that, however the surrounding support moves, the top stays pointing in the same direction. Gyroscopes are used in aircraft because they are much more accurate than a magnetic compass. The gyroscope being developed at Stanford, however, is designed to measure effects so small that it needs to be virtually perfect.

The gyroscope consists of four spinning quartz spheres about the size of ping-pong balls, which will be placed in orbit around the Earth. Since any irregularities in the balls would create effects larger than those predicted by general relativity, they have to be spherical at the scale of the atoms from which they are made, 10 000 times more spherical than the Earth. The spheres themselves are coated with a very thin, uniform layer of niobium, the best superconducting pure element. A superconductor is a material in which electrons can move without resistance. In ordinary electrical wire, the electrons move but not without friction. You can feel the effects of this resistance by picking up the electrical lead used by a power tool or heater—the lead is warm due to the effort exerted by electrons passing through the wire. In superconductors, such resistance doesn't exist and the electrons move freely. But most superconductors need to be cooled to very low temperatures before they become superconducting. In the case of the gyroscope, the whole affair will be surrounded by a shield cooled with liquid helium, which will reduce the temperature to less than 9 degrees above absolute zero. At this temperature, all electrical resistance disappears.

The superconducting nature of the niobium allows a delicate sensor called a 'superconducting quantum interference device', or SQUID, to monitor the rotation axis of each sphere. As a sphere rotates, the electrons in the niobium experience zero friction, so they lag behind the rotating metal. They stay still while the ball spins, much like coffee in a cup which is rotated suddenly. Because superconductivity is a total lack of electrical friction, the electrical current produced—the electrons flowing through the niobium layer—remains indefinitely. The electrical current is related directly to the direction of rotation of the spheres. This current creates a tiny magnetic field just large enough to be detected by a SQUID. If frame dragging is real, then it will cause the rotation axis of the spheres to move a little bit. Since the whole affair will keep pointing directly to the star with the help of the telescope, any change in the direction of the spheres' rotation axes will be detected by the SQUID.

The sensitivity of this device is astounding. You may remember that the deflection of starlight by the Sun is about one second of arc. That's the angle created by a knife-edged wedge with a thick edge of 4 millimetres and 1 kilometre long. This gyroscope has an accuracy of a millisecond of arc, the equivalent of that same knife-edged wedge gently sloping up to 4 millimetres thick over a distance of 1000 kilometres. This gyroscope is a human creation on a par with the 'Mona Lisa', the Taj Mahal or the pyramids of Egypt. In 1973, the quartz-sphere gyroscope was a '20 megabuck experiment due to fly in 1978'. Twenty-five years later, it is still on NASA's schedule and has not yet flown. The delays caused by the constant rescheduling, however, have resulted in enormous improvements. When it finally does fly, not only will it probably confirm what we already know—that general relativity is a correct description of gravitation—but it will also be the first precise test of frame dragging predicted by general relativity. This will be one of the most spectacular human achievements of the century.

Not all the effects predicted by general relativity are so subtle. We have seen how the gravity of the Sun deflects starlight. Light, like matter, follows the shortest possible path through curved space-time, and to the distant observer it looks as if the light followed a curved trajectory. If we extend this idea and imagine the light rays of a distant star passing near the Sun, each ray would be brought to a crude focus on the other side of the Sun. This effect is aptly called 'gravitational lensing', but unlike a normal glass lens, the focusing power is much stronger closer to the Sun and progressively weaker further away.

While gravitational lenses make lousy camera lenses, they do have a useful property—they magnify the amount of light passing through them. If you travelled deep enough into outer space and looked back at the Sun, you could see the light of more distant background stars magnified by the gravitational effects of the Sun. At the Sun's 'focus'—550 times the distance from the Earth to the Sun—a star exactly in line with the Sun appears as a bright ring of light around the Sun's disk. Its total brightness could be increased thousands of times. If we moved from side to side, we'd see the star's image gradually distort into a ring as the light path moved into and out of the Sun's gravitational field. This ring of light is appropriately called an Einstein ring.

We can't travel the enormous distance away from the Sun needed to look for Einstein rings. The only way these celestial mirages will appear to us is if two very distant objects are aligned with the Earth. The odds of this happening may seem absurdly remote, but keep in mind that, vast as it is, space is heavily populated with stars and galaxies. It seemed reasonable to expect that, out of the billions of objects out there in deep space, at least some of them must line up with our telescopes. Even so, it took sixty years of searching before the first gravitational lens was discovered. In 1979 a very distant object called a quasar was found, whose image had been split in two by an intervening galaxy. By carefully examining the properties of the two images, astronomers were able to prove that they were of the same object. The first gravitational lens had been found.

Since then, several gravitational lenses have been discovered, but these objects have become more than just a celestial curiosity. One of the most important problems in modern astronomy is the issue of 'dark matter'. Dark matter is the name given to the invisible matter that seems to make up the bulk of the Universe. When astronomers look out into the night sky, they're seeing the bright stars and galaxies. They began to realise earlier this century, however, that there was more mass than they could see. By carefully watching the way that stars in the Milky Way and other galaxies moved, astronomers could see that the stars were under the gravitational influence of considerable amounts of invisible matter. Look as they might, no one could see this dark matter.

In the mid-1980s, a method for looking for dark matter was suggested which involved using gravitational lenses. One of the possible sources of dark matter is a halo of dark objects, the mass of very small stars. Stars need a minimum mass in order to begin nuclear reactions and so start shining, but these massive compact halo objects, or MACHOs, lacked sufficient mass to burn. As a result, these stillborn stars simply float aimlessly and darkly among the stars. But, they have mass. If they have mass, they distort space-time. They can act as gravitational lenses.

In 1992, Project MACHO began at Mount Stromlo Observatory in Australia. Using the largest electronic detector ever made, astronomers watch the light of millions of stars in a nearby galaxy called the Large Magellanic Cloud every clear night, waiting for a MACHO to pass in front of one of those stars. When such a chance alignment occurs, the MACHO should brighten the star's light in a specific way that is characteristic of gravitational lenses. Less than a year after the project began, the scientists began detecting lensing events that indicated the presence of MACHOs. Although the jury is still out on the question of whether MACHOs make up all of the dark matter, evidence for the existence of MACHOs—detected using gravitational lensing—is being accumulated.

Gravitational lensing has also been used to measure the Hubble constant which describes the rate of expansion of the Universe. The Universe is expanding from its initial fiery birth in the Big Bang. To determine the rate of expansion, we need to know two things: the true distance to the galaxies, and how fast they're moving away. The recession velocity of a galaxy can be clocked using the same Doppler effect that police use to catch speeding drivers. In the case of galaxies, we measure the change in wavelength or frequency of the known true colours of hot hydrogen gas. This is the easy part. The hard part is to determine the distance to far-away objects.

Among the most distant objects in the Universe are quasars. These mysterious objects are incredibly bright—they have to be, to be visible from such extraordinary distances. They are thought to be the active cores of galaxies: at their heart, enormous spinning black holes are thought to be swallowing up vast quantities of gas and stars. On the way in, the matter is heated, and much is ejected again in vast beams of high-velocity particles. Quasars outshine their host galaxies like beacons which can be seen across the entire visible Universe.

Purely by chance, some quasars line up roughly with intervening 'normal' galaxies. Often the normal galaxy is so far away it is hardly visible. The quasar can shine directly through the galaxy (there is plenty of space between the stars), but its light may also take one or more paths around the galaxy, following the curved dimple in space that is the galaxy's gravitational field. You can imagine the paths that light rays travel either side of the galaxy: unless the quasar is perfectly aligned with the galaxy, the light rays will travel different distances. For every light path, you will see a different image of the quasar on either side of the galaxy. Now, quasars are not only very bright, they also vary in brightness. So, if a quasar suddenly brightens, then one image appears brighter before the other: one image from the quasar makes it to you in less time than the other, because of the difference in distance. After travelling across space for billions of years, one image arrives a few months ahead of the other.

The best example of a quasar image distorted in this way is the first one ever discovered. Called Q0957+561, the quasar appears as a double image a few arc-seconds apart and is receding from us at about two-thirds of the speed of light. In between us and the quasar is a barely discernible galaxy only a billion lightyears distant. Astronomers have been watching the quasar for ten years; they have seen the two images vary in brightness in exactly the same way, but with one image varying about 400 days behind the other. Since both images vary in exactly the same way, they must be of the same object. Relatively simple trigonometry allows astronomers to calculate how far away the quasar is: at least six billion lightyears. Knowing the distance to the quasar, coupled with a knowledge of how fast it's receding, astronomers can make a fair estimate of the Hubble constant. However, more examples are needed before an accurate measurement can be made.

Yet another use for lensing is the detection of planets around other stars. To search for extrasolar planets, you first need to find a star conveniently lined up with a distant galaxy. If the star has a system of planets around it, the planets will move in and out of alignment with the stars of the distant galaxy, causing brightness fluctuations. As yet this idea has not been tested.

While there are now many examples of gravitational lenses, none are as spectacular as those in a photograph taken by the Hubble Space Telescope in September 1994. The lens itself is a rich cluster of galaxies called Abell 2218. Lying at a distance of some three billion lightyears, the cluster's enormous gravitational field has distorted and magnified the images of the background galaxies—some three to four times further away—into several thin and faint arcs. Astronomers have now determined the approximate distances for 120 arcs, allowing them to determine the distances to galaxies fifty times fainter than is visible to ground-based telescopes.

Being able to see further into the Universe is one of the main aims of modern astronomy, since it effectively allows astronomers to peer back into time . . . closer to the beginning of the Universe. Because light from objects in space takes time to reach us, the images that astronomers receive from distant galaxies show what the galaxies were like long ago—much like an out-of-date photograph posted from a distant relative. By the time the images reach Earth, it's old news. But this is just what astronomers want, since looking at more distant, and hence younger galaxies, tells them about the evolution of the Universe. The Hubble image shows how galaxies appeared when the Universe was only a quarter of its present age. Through gravitational lensing, general relativity has become a tool for astronomers. For the first time it is being used as a cosmic looking-glass, allowing us to observe otherwise invisible objects and chart the depths of space.

The age of Newtonian flat space is over. No longer is space an empty void through which the worlds hurtle; it is not without physical properties. Space has become a tangible substance that can be bent and distorted like a rubber sheet. Einstein's most important discovery was an equation which described the elastic deformation of space. What's more, if space can be described as elastic, it should also be able to carry waves. In fact ripples in space-time are a natural outcome of general relativity. General relativity will not have come into its own until these ripples allow us to observe some of the most violent events in the history of the Universe.

CHAPTER 5
MAKING WAVES

We have seen how massive objects can bend space-time, creating a local dimple in the Universe. But what sort of objects create waves in space, and how do they do it? To answer these questions, let's first take a closer look at just what gravitational waves are like. We'll do this by comparing gravitational waves with a more familiar form of radiation: electromagnetic waves.

There is a basic difference between electromagnetic waves and gravitational waves: gravity has only one charge, and it is always positive. Electromagnetism, on the other hand, has two charges—one positive and one negative—and as a result, an electrical charge can never be accumulated in large quantities. When an electric charge builds up in a thunder cloud, for example, lightning discharges it. In this way, electricity is self-neutralising. For the same reason, atoms almost always have the same number of protons and neutrons, so they remain electrically neutral.

Gravity is quite different: it can accumulate indefinitely. Although the gravitational force on a proton is 10^{40} (forty powers of ten) times smaller than the electrical force, it is possible to accumulate so much mass that gravity overwhelms electromagnetism: for example, in the collapse of a neutron star. In fact, there are situations where the power of gravitational waves from a single source can be comparable with the total electromagnetic output power of all the stars in the Universe!

The simplest explanation of gravitational waves is that they are ripples in the curvature of space-time. This view, while accurate, is not particularly helpful for understanding their properties. However, from a basic knowledge of gravity, you can deduce some of them. One important property of gravity is this: at a distance, the gravity of a spherical body like the Earth is the same as it would be if all the mass was concentrated in a single point. In other words, the gravity at the surface of the Earth is exactly the same as if all the mass was concentrated in a single point at its centre. While the spontaneous collapse of the Earth is unlikely, such a thing can happen to a star, such as when a black hole is created. In either case, if there is no change in the gravity when the Earth or a star is squashed into a point, then there is no change in curvature of space-time and no gravitational waves.

Let's try another approach to making gravitational waves. Again, let's start with the analogy of electromagnetic waves. If you take two spheres and apply a high voltage between them, it forces a bit of positive charge to the left and a negative charge to the right. Now reverse the battery terminals. If you do this fast enough, and repeatedly, you can make the electric charge oscillate between the two spheres: the process would then create electromagnetic waves. The motion of a positive charge to the left is indistinguishable from that of a negative charge to the right, and so this type of oscillator is equivalent to having a single charge moving back and forth. This is called a dipole oscillation and is the characteristic form for electromagnetic waves.

To do the same thing with gravity, you would move a single mass back and forth. That sounds easy enough—push a mass away from you then pull it back. But no! Action equals reaction: you can only move one mass to the left if another mass recoils to the right. In everyday life, we are not aware of this because we are securely attached to the ground and the recoil of the entire planet is infinitesimal, but jumping off a small boat afloat on a lake produces a more visible effect—a recoil that can lead to waves of a different kind! Because the recoil of the two motions cancels out, we cannot have a dipole motion. In fact what we get is a quadrupole motion: the spacing between the two masses changes, but the centre of mass remains unchanged. The change in the mass distribution changes the gravity, or the space-time curvature. Since the quadrupole oscillation changes the curvature of space-time, it should produce quadrupole gravitational waves.

Since a quadrupole oscillation produces an outgoing ripple of space-time, it only seems fair that such a gravitational wave would produce the same quadrupole oscillation in a pair of suspended masses. In other words, if the motion of a mass or masses out in space produces gravitational waves, it only seems fair that those gravitational waves produce motion in a distant mass—say, a gravitational wave detector here on Earth. A passing gravitational wave would cause a pair of suspended masses to move alternately apart and together.

As you may have guessed, this is a clue to how gravitational waves can be detected. What would happen if we had four masses suspended, forming a vertical diamond figure? If a gravitational wave passed through, the horizontal masses might at first move farther apart. What do the vertical masses do? They won't move farther apart as well, since this would be like our collapsing sphere, and we've already decided that the symmetry of a collapsing sphere rules out the production—and detection—of gravitational waves. In fact, the vertical pair moves closer together. As the horizontal pair begin to move closer together again, the vertical pair start to move farther apart. On and on the cycle would go, for as long

Figure 5.1 Masses suspended in a diamond shape.

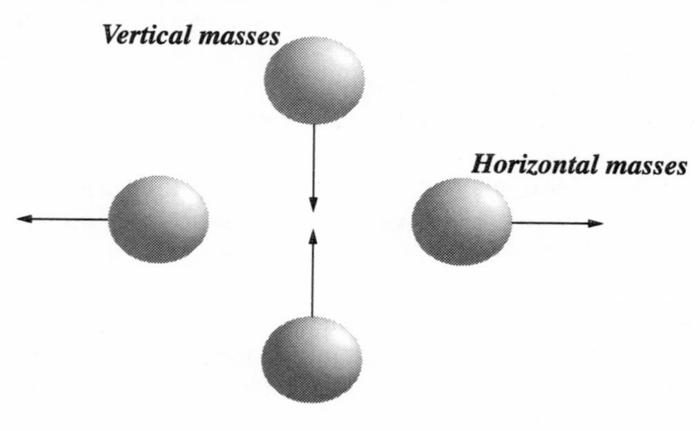

as the gravitational waves were passing through. If the four masses were replaced by a solid ring, the ring would flatten into an oval alternately in one direction and then in the opposite direction, squeezed and stretched, stretched and squeezed.

There's another aspect of gravitational waves we ought to look at before going any further, and that's polarisation. Polarisation describes the direction of oscillation in a wave. Electromagnetic waves can be polarised to oscillate in two directions, which you can think of as horizontal and vertical. Television signals are usually transmitted in a horizontal polarisation—that is why our TV antennas are flat. Polaroid sunglasses allow only one direction of oscillation to pass through, thus cutting out half of the light. In the case of gravitational waves, think about those four masses suspended in a diamond configuration. When a gravitational wave which is oscillating vertically comes through, it makes those masses move further apart and closer together. If a second wave came through hot on the heels of the first, but this time oscillating horizontally, the effect would be essentially the

Figure 5.2 Add four more masses and this is what you get.

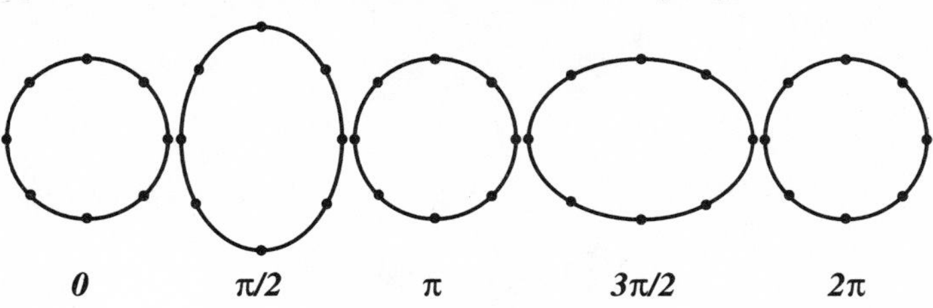

same: the vertical masses would move further apart while the horizontal masses moved closer together, and vice versa.

Let's add four more masses to this arrangement, so they create a square within the diamond. When either of the two gravitational waves passes through, the relative spacing of these new masses remains the same. Even though the vertical and horizontal masses oscillate like crazy, the diagonally placed masses remain stationary. The only way to make them move is to rotate the gravitational wave by 45 degrees, so that it's now alternately squeezing and stretching the diagonal masses. But wait, now the vertical and horizontal pair are stationary! Clearly, gravitational waves have two polarisations 45 degrees apart.

From a simple starting point, we have deduced the basic structure of gravitational waves: quadrupole oscillation of masses creates waves, and these waves in turn create quadrupole oscillation in masses. The next question to ask is: how much energy does it take to produce a gravitational wave?

The thing to remember about space is that it is very stiff. What we mean by this is that it is very difficult to create any sort of vibration in space. If you were to stretch a long rubber band and then pluck it, you'd hear and see the vibrations—ripples—travelling along its length. If you tried to do the same thing with something

Figure 5.3 Gravitational waves have two polarisations 45° apart.

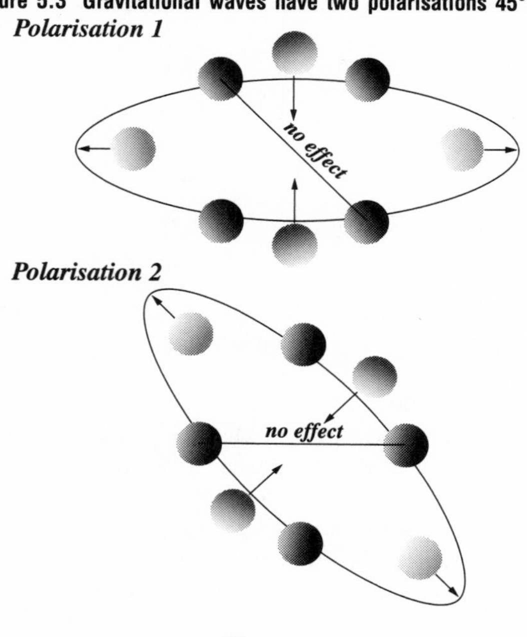

much stiffer, say a railway line, all you would do is bruise yourself trying to get it to vibrate. You know what to do. Use a good-size sledge-hammer and give the line a good bash. You don't have to be a physicist to understand what's happening here: the wave power you inject into the line depends on the mass of the hammer, the frequency with which you hit it, and it also depends inversely on the stiffness of the line (the stiffer it is, the less wave energy generated).

Now, you can think of the two masses we used to create gravitational waves as the equivalent of a sledge-hammer. In this case, the gravitational wave power that you produce depends on the masses, their frequency of oscillation, and on the elasticity of space-time. If the masses are close together, you would expect the gravity of the pair to be similar to the gravity of a single mass. Hence small changes in spacing between the masses will have less effect on space-time curvature. So the gravitational wave power produced should also increase with the spacing of the masses, and, of course, with the amplitude of their oscillation.

Putting all this together, we can say that:

$$\text{Gravitational wave power} = \frac{\text{mass} \times \text{spacing} \times \text{amplitude} \times \text{frequency}}{\text{stiffness of space–time}}$$

This result shows the main feature of gravitational waves: they are weak, not because there aren't some very violent processes in the cosmos, but because the wave power is divided by the enormous elastic stiffness of space-time (about 10^{43}).

The most effective source of gravitational waves is one where the spacing between the masses changes from nearly zero to some maximum value. Two masses joined by a spring would oscillate in this way, but two masses joined by a rigid rod and spun like a band leader's baton is much better. Viewed side-on, such a rotating dumbbell appears to expand and contract from maximum to zero. There's a problem, however. There's a limit to how fast something can rotate before it flies apart under centrifugal force. Even a steel ball-bearing will explode if it rotates faster than a million times a second. This rotational speed limit makes things look quite depressing for anyone planning to send signals via gravitational waves.

Undaunted, let's try to devise the best possible gravitational wave source. We want to take two large, joined masses and rotate them as fast as possible. We've seen, though, that the bar that joins them will break at high speeds, and so it's better to use a solid elongated mass. Let's be ambitious and use 1000 tonnes of steel, or even better a 10 000-tonne nuclear submarine. If we mounted the submarine on a turntable and rotated it to just short of its breaking point—say 10 times a second—how much gravitational wave energy would we produce? Sadly, not very much at all: about 10^{-24} watts. To put this into perspective, the smallest ant walking fast up a wall uses 10^{-7} watts, a billion billion times more energy than the gravitational waves produced by our rotating submarine. And the ant's efforts are 10 billion times smaller than an average family car.

Such small amounts of energy aren't new to astronomers. Radio astronomers, for example, are quite familiar with the feeble energy they find in studying the radio universe. Yet radio telescopes have one major advantage: they can collect almost all of the energy in the incoming radio waves. A gravitational wave detector on the scale of the submarine, however, could detect only a tiny fraction of the energy originally transmitted (even if all the gravitational waves passed by it).

But there are much larger oscillating masses out there in space. Binary stars, for example, behave like sun-sized dumbbells. They consist of close-spaced pairs of stars whizzing around each other. Does the existence of these objects improve our chances of detecting gravitational waves? To calculate the strength of an astrophysical source of gravitational waves, all you have to do is put the numbers into Einstein's quadrupole formula, much like the one we derived above. We could apply it to ordinary binary star systems, where the stars orbit each other as fast as once a day. In our Galaxy there are millions of such systems which, as a whole, should produce a background of random ripples in space-time at a very low frequency of one cycle per day. Each half of the binary system creates a full in-and-out motion of the masses, as seen edge on, so the gravitational wave frequency is double the orbital period. This means that the gravitational wave frequency is in the range of one cycle every few hours to one cycle every few days. Each individual source is very weak and even the combined effect of all of them would be completely undetectable by Earth-based detectors, where disturbances such as temperature variations and tides cause relatively enormous variations on the same time scale as the signal. As a result, gravitational waves from these sources are completely swamped, although, as we will see later, they may be detected from space.

This depressing situation was understood by Einstein. From the 1920s to the 1960s, relativists believed that gravitational waves were of academic interest only. In 1961, the distinguished British astrophysicist Herman Bondi wrote: 'Gravitational waves are not only unfamiliar, even by name, but are distinctly unlikely to be observed.'

The only solutions were to build better detectors, or to come up with a better source of gravitational waves. In fact, both of these needs would be fulfilled in the 1960s. Decades earlier, predictions had been made for bizarre objects of unbelievable density: neutron stars and black holes. Neutron stars are a superdense ball of neutrons 20 to 30 kilometres in diameter, which yet contain the mass of an entire star. With such strong gravity, the escape velocity from a neutron star is very high—about 10 per cent the speed of light. In isolation, neutron stars are unlikely to be strong sources of gravitational waves, despite the fact that some of them rotate hundreds of times each second. However, they are often in mutual orbit either with an 'ordinary' star or with another neutron star, closely fulfilling the needs of our rotating mass source of gravitational waves. Not only do neutron stars rotate extremely fast, but they are also capable of very fast orbital motions. To cap it off, they are exceptionally good clocks. In short, neutron stars are a relativist's dream come true. Black holes are a close relative of pulsars, and their formation is also believed to be a possible source of gravitational waves. These strange objects are the result of the entire mass of a star (or more) collapsing into a single point of infinite density. Around it is a region of space from which not even light can escape.

Before we look at how neutron stars and black holes can produce gravitational waves and, in the case of neutron stars, prove the existence of gravitational waves, we'll take a detour and explore their origins—itself a possible source of gravitational waves.

Chapter 6
Stars that go Bang in the Night

The massive star wandered through the galaxy as it had for millions of years. Although its 10-million-year existence was a mere blink compared with the 9000-million-year lifetime of a small star like the Sun, to the massive star it was all the time it had. It had shone extremely brightly in the night sky, and its lifetime was correspondingly shorter, and its death was going to be spectacular. Not only would the event outshine all its neighbouring stars in the electromagnetic spectrum, but it might send gravitational waves rippling across the Galaxy.

Like all stars, the massive star had begun by condensing within a large cloud of hydrogen, along with dozens of others. The cloud broke up into fragments of various sizes, including a very large one. The fragments became 'protostars', slowly contracting spheres of hydrogen that glowed dimly in space. Like the others, the more massive protostar contracted; the temperature in the centre rose. At last the temperature reached some 10 million degrees Celsius, hot enough for the hydrogen atoms to fuse into helium, giving off tremendous amounts of energy in the process. The star was born. The thermal pressure from the nuclear reactions going on inside the star nicely balanced the crushing weight of the star itself. But its tremendous bulk weighed heavily on the core, causing the hydrogen fuel to burn at a furious rate. A mere 20 million years after the star had emerged from its stellar nursery, it faced a crisis as its store of hydrogen fuel began to run out.

As the hydrogen fused to form helium, the heavier helium sank to the centre of the star. The star now had a helium core surrounded by a shell of hydrogen which continued to burn. After a time, with no nuclear reactions to sustain it, the helium core began to collapse in on itself. The outer layers of the star, no longer bound by the bulk of the collapsing star, expanded and cooled. The star became a red giant. As the helium core collapsed, it became denser and hotter. Eventually the temperature in the core rose to 50 million degrees, hot enough to trigger the fusion of helium into carbon, restoring the star's source of radiant energy. It had been given a reprieve. With time, the core turned to carbon and the star once again faced a turning point. The cycle repeated and the carbon was converted to neon, and within a thousand years the neon began to burn to silicon. Within a year, the star's core consisted of silicon. The collapsing core rose to a temperature of 400 million degrees, and began to burn the silicon to iron. The process was complete within a few days.

Here at last was the final breath the star would take in its present incarnation. All of the previous reactions converted lighter elements into heavier elements, producing energy that kept the star shining. Iron cannot be converted into a heavier element without using energy. With nothing left to burn, the star's core collapsed catastrophically. The central core collapse took a few milliseconds, maybe halting only when it was a 30-kilometre-diameter neutron star, or maybe falling all the way to a black hole. Some of the inward falling matter bounced off the collapsing core and was blown outwards in a titanic explosion. In the milliseconds of the collapse, a vast pulse of gravitational waves, comparable in power to all the stars in the universe, was created, and expanded outwards at the speed of light. For a few weeks thereafter the dying star shone brighter than a galaxy of 100 000 million suns in visible light.

This dramatic chain of events is typical of one of the most violent processes in the Universe: a supernova. It is one of the best sources of gravitational waves.

The death of massive stars as supernovae occurs at irregular intervals throughout the Milky Way and other galaxies. With so much light produced by a single event, the question arises: has anybody ever seen one from Earth? During the past 2000 years there have been about ten occasions when civilisations have witnessed a supernova. Perhaps the brightest of all occurred on 1 May in the year AD 1006. Chroniclers from China, Japan, Arabia and Europe all recorded the event, testifying to its impact. The following are just a few of the records of the event.[1]

From the 32 volumes of the German monastic chronicles called the *Monumenta Germaniae Historica Scriptores* come numerous accounts:[2]

> 1006. In the 25th year of our master Pandolphus and the 19th year of our master Landolphus, his son, a very brilliant [*clarissima*] star shone forth, and there was a great drought for three months.[3]

> A new star of unusual size appeared, glittering in aspect, and dazzling the eyes, causing alarm. In a wonderful manner this was sometimes contracted, sometimes diffused, and moreover, sometimes extinguished. It was seen likewise for three months in the inmost limits of the south beyond all the constellations which are seen in the sky.[4]

> There was a very great famine and a comet appeared for a long time.[5]

> Three years after the king [Henry II] was raised to the throne of the kingdom, a comet with a horrible appearance was seen in the southern part of the sky, emitting flames this way and that.[6]

In Europe, the supernova was very low on the horizon, yet it was clearly impressive. Despite its 'wonderful' and 'brilliant' appearance, it was generally taken as a bad omen, being seen as an interference with God's firmament, which was supposed to be fixed and unchanging. The supernova brought with it alarm, droughts and famine.

> 1006. In this year was seen in the sky a burning star like a torch which is called a comet.[7]

Chinese records are more accurate and show a more positive reaction to the supernova. The *History of the Sung Dynasty* states:

> On the fifth day of the fourth month of the third year of the Ching-te reign, a *chou-po* star was seen. It appeared to the south of Ti, and one degree west of the Qan [constellation of Lupus, 40° south]. Its form was like the half-moon with pointed rays shining so brightly that one could see things clearly.[8]

The best description of the supernova discovery is found in a book presented to the Sung Emperor in 1044:

> The Director of the Astronomical Bureau reported that at the first watch of the night on the 2nd day of the fourth month, a large star, yellow in colour, appeared to the east of K'u-lou at the west of Ch'i-kuan. Its brightness had gradually increased. It was found in the third degree east of Ti. . . . The star later increased in brightness. According to the Star Manuals there are four types of auspicious stars. One of these is called *chou-po*; it is yellow and resplendent and forbodes great prosperity to the state over which it appears.[9]

From the personal diary of a thirteenth-century poet-courtier in Japan comes another report which confirms the discovery in the early evening of 1 May: '3rd year of the Kanko reign period of Ichiijo In, 4th month, 2nd day, Kuei-yu. After nightfall, within Ch'i-kuan there was a large guest star. It was like Mars and it was bright and scintillating.'[10]

Another report from China emphasises how disturbing the supernova was for ordinary people, and how a clever official, Chou K'o-ming, brought the supernova to everyone's advantage—a holiday celebration for the people and a promotion for himself.

> During the third year of the Ching-te reign period a large star appeared at the west of Ti. No one could determine its significance. Some suggested that it was a baleful star which portended warfare and illfortune. At that time, K'o-ming was away on a mission. On his return he urgently requested permission to reply to these suggestions. He said, 'I have checked the star manuals. The interpretation is that the star should be called a *chou-po* star, which is yellow in colour and resplendent in its light. The country where it is visible will prosper greatly, for it is an auspicious star.' Then he went on to say, 'On my way back I heard that people inside and outside the court were quite disturbed. I humbly suggest that the civil and military officials be permitted to celebrate in order to set the Empire's mind at rest.' The Emperor approved . . . and promoted him to the post of Librarian and Escort to the Crown Prince.[11]

From further south, in Arabia, the new star was higher in the sky. Quantitative factual records demonstrate the much higher numeracy of the Arabs at that time. Yet these were still combined with ominous interpretations similar to those in Europe.

> Year 396. Among the incidents of that year a large star

> similar to Venus in size and brightness glittered to the left of *qibla*. Its rays on Earth were like the rays of the Moon.[12]

> The spectacle was a large circular body, 2.5 to 3 times as large as Venus. The sky was shining because of its light. The intensity of its light was a little more than a quarter of that of moonlight. It remained where it was and it moved daily with its zodiacal sign until the Sun was in sextile with it [i.e. 60 degrees away] in Virgo, when it disappeared . . . At the time when the spectacle appeared, calamity and destruction occurred which lasted for many years afterwards.[13]

If we look in the direction described in the records, we find a classical supernova remnant: a roughly spherical cloud of radio emission, similar in apparent size to the full Moon, but corresponding in physical size of at least 10 lightyears. This huge structure results from the outer layers of the star which have been ejected at thousands of kilometres per second.

It is difficult to estimate the size and brightness of sharp points of light. Light scattering in the air always increases in apparent size, and the saturation of the retina distorts our perception of its brightness. Comments like 'one could see things clearly' and 'the sky was shining' best illustrate the brightness of the event. In bright moonlight you can see trees and landscape quite clearly and you can often see the bright halos of scattered moonlight making the sky shine. The supernova must have been like a streetlight in the sky, dramatically brightening moonless nights. In Australia, where it was overhead, it must have been truly spectacular. We wonder if there are any oral records passed down by the Aboriginal people.

The spectacular brightness of the supernova arises mainly from the explosive expansion of a star. The bright, hot, outer parts of the star are blown outwards so fast that the star diameter increases at about a million kilometres every minute. Within a few hours, if the Sun were a supernova, it would engulf the Earth. Within a few days, it would be as large as the solar system. Very hot material from deep within the star would have been exposed so that a distant observer would see the star rapidly brightening until its light output was comparable to an entire galaxy of a hundred billion stars.

Just 48 years after the supernova of AD 1006, there was another comparable spectacle. This one was carefully recorded by Chinese and Japanese astronomers. A bright 'guest star' had appeared on 4 July 1054 in the constellation of Taurus. It was visible in daylight, and continued to be visible for almost two years. Today at the site of this supernova we see a spectacular nebula—the Crab Nebula—and at its core is the Crab pulsar flashing a double pulse 30 times a second at almost every wavelength of the electromagnetic spectrum from radio to gamma rays.

Earlier in this century, astronomers had realised that the nebula was expanding. Extrapolating backwards, they predicted that there should have been a supernova in the eleventh century in Taurus. In 1942 Mayall and Oort proved definitively that the Crab Nebula was the remnant of the spectacular 1054 supernova. Finally, the discovery of the Crab pulsar in 1968 confirmed the link between neutron stars and supernovae. Yet do not believe that all supernovae create neutron stars: in many other instances, no such clear link exists. The whole picture of how pulsars are born and what happens in supernovae is still confused.

If the historical supernovae represented the total number of supernovae to have occurred in our Galaxy during the last millennium, this would imply a supernova rate in the Milky Way of roughly one supernova every 140 years. However, all but one of these supernovae occurred—in our corner of the Milky Way—within about

12 000 lightyears of the Earth. The distance to the centre of our Galaxy is at least double this distance. Presumably supernovae also took place in other parts of the Galaxy; these were simply too far away and hidden by dust.

We live in a sparsely populated corner of the Milky Way. You can get a better idea of what goes on in galaxies by looking at lots of distant galaxies. By looking at supernovae in other galaxies, and averaging over the galaxies, you find that supernovae occur more often towards the centres of galaxies. This ties in with the denser stellar populations found there. There is a surprise in this, however. A kind of of supernova known simply as Type I is believed to occur only among old stars which are found towards the centres of galaxies, and yet Type I supernovae are found randomly distributed throughout the total numbers of supernovae in galaxies. Anyway, by looking at how often supernovae occur in other galaxies, and assuming our galaxy is typical, the historical record of supernovae that have occurred in our outlying corner of the Galaxy implies that the total number of supernovae in the Milky Way could have been 100 in the last thousand years, and perhaps one every eight years.

Usually, precise numbers aren't important in astronomy, but when it comes to supernovae and searching for gravitational waves it is important to know whether they occur every twenty-five years or every five years in the Galaxy. The reason is that, while it's reasonable to leave a gravitational wave detector running for five years with a good chance at detection, a supernova every twenty-five years sounds like a long shot, and who wants to wait that long?

Of course, the observed rate of historical supernovae may be anomalously high. With so few events, statistics can play havoc. Most estimates of the supernova rate in external galaxies are much lower, from one every thirty years to one every hundred years. The trouble with these estimates is that many supernovae are bleached out in the photographic image of a galaxy, especially near the centre, and dim ones are missed. This bleaching happens because the central regions of galaxies are so bright that the photographic image is almost totally white and a supernova will not show up.

The high rate of supernovae implied from the historical records has one big problem: if supernovae actually occur every eight years in our Galaxy, how come radio astronomers have not detected some very young remnants—baby versions of the Crab nebula? The problem may be in one key assumption: that supernovae are evenly distributed around galaxies. We may live in a special part of the Galaxy, one where stars are forming strongly in our local spiral arm and where big stars evolve quickly to the supernova stage. As we saw at the beginning of this chapter, a star twenty times as massive as the Sun lasts a mere 10 million years before it explodes as a supernova, while the Sun has been around for five hundred times longer. The historical supernovae may represent a much larger fraction of the total number in our Galaxy than you would expect if we were not in a lucky location where we get a prime view of a supernova-rich neighbourhood.

About fifty supernovae are observed every year in distant galaxies. They are very faint and certainly not visible to the naked eye. They're often discovered by accident on photographs where they appear as bright, point-like objects, almost as bright as the galaxy in which they occur.

The Reverend Robert Evans in Australia has developed an uncanny ability to discover supernovae simply by monitoring a large number of galaxies night after night with a small telescope. The galaxies look like fuzzy blobs, and most of them include superimposed foreground stars in our own Galaxy. Evans is able to remember the star patterns well enough to recognise any new stars in the field of view surrounding the galaxy he is observing. Such new stars inevitably turn out to be supernovae in the external galaxies. Given Evans's religious status, you can imagine the jokes that are made about him being privy to advance warning!

Astronomers have recently tried to take the drudgery out of supernova searching by making robot versions of Evans. The idea is to use a telescope with an electronic CCD (charge-coupled device, a light-sensitive electronic detector, a kind of 'electronic film') camera programmed to search galaxy after galaxy for supernovae. The images that the telescope creates are automatically compared with library images: any difference in the number and position of stars sets off an alarm. Carl Pennypacker's group at the University of California at Berkeley was the first to operate such a telescope successfully. Following his lead, a group in Perth built a system on a shoestring budget using only PC computers. Both systems have successfully detected supernovae, but so far with not much better success than Evans. Very few supernovae have been detected independently by more than one search. This tells us that our supernova searches are detecting only a small fraction of all the supernovae that occur in our corner of the Universe.

Unfortunately, the distant supernovae do not completely clarify the situation. We can only observe the supernova's light curve (a plot of how the brightness varies over time) and the spectrum of its light. From this data we can tell that supernovae come in different flavours, but they are too far away to be able to see what remains after the supernova explosion (the supernova remnant), or to see if a neutron star or black hole has been left behind.

Here again we have to resort to the small historical collection of supernovae and the supernova remnants in our galaxy that can be more-or-less accurately dated. One of the supernova remnants indicates that the explosion occurred almost 350 years ago. All that's visible today is a huge, rapidly expanding, hollow spherical shell. As the high-velocity gas collides with the tenuous interstellar medium, it creates radio emission, allowing radio telescopes to produce an image. In fact, it is one of the brightest radio sources in the sky. By carefully observing the expansion rate, the expanding shell's size can be traced backwards and the date of the original explosion estimated. The supernova occurred about the year 1657.

In that era, astronomers such as Flamsteed should have been able to study the event in detail, but it seems to have been overlooked. This is puzzling since, although there is some obscuring dust which would have dimmed the event somewhat, it should still have been easily visible. The Russian astrophysicist Schlovsky suggested that there was no visible-light outburst from the supernova because the original star collapsed directly to form a black hole. However, if this had happened and all the material collapsed inward, there would be no radio remnant today. Perhaps the lack of observation in 1657 was a combination of dust, poor weather, bad luck (nobody looked in the right direction) and lack of recording (if they did notice it, they didn't write it down).

The second historical supernova that was not recorded is MSH–103. This occurred in the southern sky and could have been seen only by civilisations without writing. There is every chance that it was seen in Australia, South Africa or South America, but there are no records to prove it.

The list of historical supernovae ends in 1604, or if we include the 1657 remnant, about fifty years later. There has been no supernova visible to the naked eye in our galaxy for nearly 400 years. This statistical fluctuation seems very unfair! The only compensation was the supernova 1987A which occurred in the Large Magellanic Cloud, our nearest neighbouring galaxy, 150 000 light-years away, on 23 February 1987. This was visible to the naked eye as a faint star in the southern sky (northerners missed out!) and is part of the complex picture we will explore in the next chapter.

Are supernovae sources of gravitational waves? In the next chapter, we'll be talking more about neutron stars, one possible by-product of a supernova. Although these city-sized spheres rotate at fantastic speeds by everyday standards, they're rotating slower than would be expected if they are the result of a collapsing giant star, as most astronomers believe. The fact is, if a giant star collapses, its rotation increases to amazing speeds as it does so, much as a spinning ice-skater can spin faster by drawing her arms closer to her body. The problem is, by the time you take a giant star and compact it into a sphere 20 kilometres across, it is spinning even faster than pulsars do, in fact, faster than is possible by any stretch of the imagination—faster than the speed of light. The question then arises: where does all that rotational energy go?

To answer the question, think of a star collapsing and spinning faster and faster. All stars and planets bulge at the equator as they spin on their axes. A star rotating faster as it collapses will bulge even more, becoming flatter like a pumpkin. But this can't go on forever. There are three possibilities: (a) it may form a ring, which itself will be unstable and break into several pieces; (b) it may develop two opposite bulges, which grow until it becomes two separate objects resembling two stars just touching, a process called bar-mode instability; or (c) it may remain cylindrically symmetrical and shed matter at its rapidly rotating equator.

The first two scenarios provide good sources of gravitational waves. The gravitational waves emitted by these objects carry away angular momentum, allowing the objects to collapse into a neutron star. Whether gravitational waves can carry off enough rotational energy to leave the pulsars rotating at their observed rates is a matter of serious research.

So, supernovae may be likely sources of gravitational waves and they may go off every ten years, or every hundred years in a large galaxy like the Milky Way. As we look further afield, of course, we encompass more and more galaxies and more and more supernovae. In every direction astronomers are seeing the sky thick with galaxies. If there are 10 000 galaxies within 30 million lightyears, then there are 10 million galaxies within 300 million lightyears and 10 billion galaxies within 3000 million lightyears, because the volume increases as the cube of distance. Ten billion galaxies each hosting a supernova every thirty years corresponds to ten supernovae every second. Within the observable Universe, supernovae should be exploding at a rate of anything up to 1000 per second.

Most of these supernovae are very, very distant, and the gravitational waves from a single event are very weak. However, when there is a nearly continuous signal it is possible to use correlation methods to prove that it is present, even though the single events cannot be detected individually. To do this you use two detectors, and combine the signals in such a way that the background noise of the instruments cancels out, while only the very weak signal common to both instruments gets accumulated. By using such averaging, the combined effect of all the supernovae in the observable Universe could be quite easily detected when new, big detectors come on line.

This supernova signal will be very interesting. In gravitational waves, supernovae probably sound a bit like popcorn going off in a popcorn maker. The cosmological supernova signal probably sounds like a football stadium filled with popcorn makers. Sitting in the middle you will hear close but infrequent loud pops, fading to a continuous roar in the background. The precise way it fades from pops to a roar will tell us how the rate of supernovae has changed from the time when galaxies were born to the present when galaxies have settled down to old, sedate spirals like the Milky Way. Many people think that the rate of supernovae in the early universe was a hundred times greater than it is today. In this case we should hear this aspect of cosmic evolution in the sound of popcorn noise.

Notes

1 Richard Stevenson, David Clark and David Crawford, *Monthly Notices of the Royal Astronomical Society*, 180, 567 (1977).
2 G.H. Pertz (ed.), *Monumenta Germaniae Historica Scriptores*, 32 volumes, Hahn, Hanover (1826–).
3 From 'Annales Beneventani' (Annals of the Monastery of St Sophie, Benevento), in *Monumenta Germaniae Historica Scriptores*.
4 From 'Annales Sangallenses Maiores' from the chronicle of the Benedictine Monastery of St Gallen, in *Monumenta Germaniae Historica Scriptores*.
5 'Annales Laubienis' and 'Annales Leodiensis' (Belgium), in *Monumenta Germaniae Historica Scriptores*.
6 'Alpertus de Diversitate Temporum', Lib. I, in *Monumenta Germaniae Historica Scriptores*.
7 'Annales Mosomagenses' (Annals of Mousson), in *Monumenta Germaniae Historica Scriptores*.
8 *History of the Sung Dynasty*, chapter 56.
9 Ch'ing-li-kuo-chao-hui-yao, AD 1044.
10 Meigetsuki, Personal Diary of Fujiwara Sadaie, vol. 3
11 The Sung-shih biographies: Biography of Chou K'o-ming, AD 954–1017.
12 Kitabal-Muntazam, Ibn al-Jawzi, who died in AD 1200.
13 Ali ibn Ridwan, *Commentary on the 'Tetrabiblos' of Ptolemy*.

CHAPTER 7

THE COMING OF THE PULSARS

Central to general relativity is the relationship between gravity and time. Einstein developed his theory by considering imaginary observers with perfect clocks. These *gedanken* (thought) experiments always seemed frustratingly beyond reach. To fully investigate general relativity we need strong gravity, high velocity and perfect clocks. Strong gravity means gravity or curvature strong enough that the deflection of light is a few degrees, not just the few arc-seconds that Eddington measured (Chapter 4). Strong gravity also means that the escape velocity should be a substantial fraction of the speed of light—say a few per cent rather than the few hundred-thousandths we experience on Earth. High velocities mean velocities comparable with the speed of light. Although in the solar system we have access only to relatively low speeds and low gravity, we do have pretty good clocks which are capable of measuring changes in time to the order of a few parts in 10^{16}. Out there in space, however, there are natural laboratories that incorporate the fantastic physical properties of gravity and speed, and as an added bonus are fantastic clocks. These natural laboratories for general relativity are called pulsars.

Pulsars are neutron stars—the rapidly spinning remains of massive stars—which emit beams of radiation like celestial lighthouses. Pulsars were discovered in 1968, but the idea of neutron stars goes back to before World War II.

In 1932, Ernest Rutherford was director of the Cavendish Laboratory at the University of Cambridge. Working with Rutherford was his former student James Chadwick. Rutherford had discovered the structure of the atom as a small, dense nucleus surrounded by clouds of electrons. But Chadwick discovered the neutron, a small, chargeless particle about the same mass as a proton found in the nuclei of atoms. The discovery was to lead the way for two important events—the discovery of neutron stars, and the first atomic bomb.

The Russian physicist Lev Landau was working in Copenhagen at the time he heard of Chadwick's discovery. As the story goes, it was only a matter of hours before Landau came up with the idea of a star made entirely of neutrons. Such a 'neutron star' would be very small, very dense, and very faint. Also working on the problem was Robert Oppenheimer and his student, George Volkoff, at Berkeley. They built on the brilliant work of Subrahmanyan Chandrasekar who, two years earlier at the ripe old age of nineteen, used the latest ideas in quantum mechanics to explain the mechanism by which matter a million times denser than water could exist in white dwarf stars. Landau and Oppenheimer extended these ideas and showed that nuclear matter, a million billion times denser than water, could exist in neutron stars. Having explained this extreme state of matter, both men became immersed in some of the extreme events of this century. Oppenheimer went on to develop the United States' first atomic bomb, and then to suffer the McCarthyist purges. In his own country, Landau was imprisoned and nearly died during the Stalinist purges.

The theory of neutron stars was independently put forward by Fritz Zwicky and Walter Baade in 1934. In an inspired guess, they suggested that neutron stars would be the by-product of supernovae, the end result of massive stars that had used up their supply of nuclear fuel. The outward rushing of the outer layers of the star was simply the by-product of the collapsing stellar core. At least as massive as the Sun, the core would shrink to only about 30 kilometres in diameter. At this density, the electrons and protons would fuse to form neutrons . . . and a thimbleful of the stuff would weigh a hundred million tonnes.

The process where the electrons and protons of normal matter merge to form neutrons is called 'inverse beta decay'. The neutron star is prevented from collapsing even further by a force called 'degeneracy pressure'. This force is described by the Pauli Exclusion Principle which says that no particle—electron, proton or neutron—can have the same energy state. Since no two neutrons can have the same velocity, all the neutrons in a star are 'stacked up' in velocity, and quantisation means that these are separate discrete values. The more neutrons in a star, the higher the stack of velocity values. In other words, the more mass there is, the faster the fastest neutron. The outward pressure arises directly from the velocity of the neutrons, just as the pressure in a balloon arises from the velocity of gas atoms bouncing off the balloon's inside surface. In a neutron star, just as in the atmosphere of the Earth, the velocity of the particles is balanced against the pull of gravity. On Earth, this determines the thickness and density of the atmosphere. In the neutron star, the pressure prevents the star from collapsing into a black hole.

Chandrasekar used these ideas, but applied them to electrons. Electron degeneracy pressure is 1800 times weaker than neutron degeneracy pressure, because electrons are 1800 times lighter than neutrons. Electron degeneracy allows the existence of white dwarf stars as first proved by Chandrasekar. Like neutron stars, white dwarfs can exist in only a small range of masses, from about 0.1 of a solar mass to 1.44 solar masses. Electron degenerate matter consists of atomic nuclei embedded in a sea of electrons: the electrons are no longer bound to individual atoms. If a white dwarf lost some of its mass so that it was less than about one-tenth of the sun's mass, it would expand and revert to a 'normal star' a hundred times larger.

White dwarfs and neutron stars represent the two possible states of degenerate matter. Each collapses above a critical mass: the white dwarf to a neutron star, and the neutron star to a black hole. Both observation and theory agree that the critical mass for a white dwarf is 1.44 solar masses. For neutron stars the number is less certain, but is somewhere between 2 and 3 solar masses. If more mass is added to a neutron star at this critical mass, the velocities of the neutrons begin to rise even further. There's a limit to how much mass you can add, however, due to the limit on how fast the neutrons can move. If you add so much mass that the speed of the fastest neutron is close to the speed of light—and remember that the speed of light is an absolute—then adding still more mass simply increases the gravity of the neutron star without increasing the outward pressure. This can only lead to one thing: further collapse into the most bizarre object in nature, the black hole singularity.

Both white dwarfs and neutron stars have the same remarkable property: add mass and they shrink. As you add mass, the density and gravity increases. At the critical mass, they collapse. But how could a neutron star form? What fantastic physical process could lead to such a remarkable object?

One way neutron stars are believed to form is from the collapse of massive stars, in other words, during supernova explosions. We've already seen how stars live their lives busily converting lighter elements into heavier ones, producing copious amounts of energy in the process. This energy keeps the star inflated against the pull of gravity. When the star runs out of fuel, it collapses under its own weight. When this happens to a star, it has one of four destinies: (a) explode totally; (b) collapse to form a white dwarf; (c) collapse to form a neutron star; or (d) go all the way and become a black hole.

Oppenheimer and Volkoff worked out the possible structure of neutron stars in 1939. But the neutron star remained a neutral, inert entity in the minds of astronomers. It wasn't until 1967 that the Italian Franco Pacini suggested that neutron stars could be rapidly spinning, highly magnetic objects. The fact remained, however, that even if neutron stars did exist, they were unobservable, and so they were all but ignored. Ignored, that is, until the results of an accidental observation made in late 1967 filtered through to the astronomical community.

In that year, a young PhD student, Jocelyn Bell, was wading through kilometres of paper chart. The traces on the chart carried information on scintillation, the interplanetary equivalent of the terrestrial atmospheric effect that causes the stars to twinkle. The observing program was being run by Antony Hewish, using a radio telescope specifically designed and built for the project. It was Bell's job to identify scintillating sources, plot them on a map, and distinguish between true sources and man-made interference. But there was an occasional signal, 'scruff' as Bell called it, that didn't look like either scintillation or interference. The same scruff showed up at the same declination and right ascension each time, in the constellation of Vulpecula.

Closer inspection of the source showed that it was flashing every 1.33 seconds. At the time there were no known natural objects that could produce such regular and fast pulsations, so the immediate reaction was that the signals were man-made. However, when the astronomers realised that the source kept sidereal time—that it moved with the stars rather than according to Earth's

24-hour cycle—they were forced to conclude that the source was celestial rather than terrestrial.

The temptation to think of the source as anything but natural was so great that at one stage the team entertained the idea that it was the result of extraterrestrial intelligence. But there were problems with the idea. First of all, the sender of the message would have to be transmitting from a planet orbiting a star. The pulsations would be Doppler-shifted as the planet approached and receded during its orbit around the star. No such Doppler effect could be found. The idea of extraterrestrial intelligence was finally put to rest when Bell discovered a second pulsating source in Leo. It was unlikely that two civilisations would be transmitting toward Earth on the same frequency from two different parts of the sky.

So what was it? The initial reactions by the astronomical community were along the lines of a pulsating star (hence the name 'pulsar') similar to Cepheid variable stars that rhythmically expand and contract. But the signals were pulsing too fast to be explained by this phenomenon. Oscillation didn't work, either, even if the stars were small and dense like white dwarfs. Even the hypothetical neutron star didn't work—according to calculations, they oscillated too fast. Nothing known could pulsate at the observed speed.

Thomas Gold was teaching at Cornell University in the US at the time pulsars were discovered. He suggested that the radio pulses rather than being produced by pulsation or oscillation, might be produced by rotation. After all, rotating bodies are among the most regular phenomena in the universe. The Sun rotates every 28 days, the Earth every day, Jupiter every ten hours . . . with monotonous regularity. But what could be rotating once every second?

As we saw with the collapse of a massive star prior to a supernova explosion, if a rotating object collapses under its own weight it spins faster. White dwarfs are collapsed stars that rotate with periods measured in minutes. If pulsars were collapsed stars, the rotation rates required by pulsars could be explained: a once slowly rotating massive star would speed up to spin at fantastic speeds.

But there was a problem with this idea. If an object were rotating once every second, the centrifugal force would overcome the gravitational force and the object would fly apart. The only way such a rapidly rotating object could remain intact is if it were massive enough to overcome the centrifugal force. Neutron stars, Gold suggested, were able to do this nicely. The gravity associated with neutron stars is more than enough to hold them together, while the rotation rate explained the pulse period.

But what could be causing the radio pulses themselves? The rotating magnetised star is like a dynamo—it has an intense magnetic field and it creates an intense electric field as it rotates. Electrons are accelerated in the intense electric field. They spiral around the magnetic field lines, and produce beams of radio waves. The process is called synchrotron radiation and is widely used on Earth to create ultraviolet light and X-rays in large-particle accelerators. Because the neutron star's magnetic axis is offset from its rotational axis, the beams sweep space like the beam from a lighthouse. When the beam flashes across the Earth, radio pulses are detected.

The discovery and subsequent explanation of pulsars sparked many searches for more of them. Surveys were needed to determine the population and distribution of pulsars in the disk of the Galaxy, as well as in the globular clusters that inhabit the halo of the Galaxy. One of the first surveys was conducted at the Molonglo Radio Observatory in Australia, during which 31 new pulsars were discovered. This was followed by the Jodrell Bank survey in 1972 (39 new pulsars) and the University of

Massachusetts Arecibo Observatory survey in 1975 (40 new pulsars). Nine years after their discovery, 149 pulsars had been discovered. But the most successful survey ever performed took place in 1977.

The 1977 Molonglo Pulsar Survey was carried out by Dick Manchester and his colleagues from the Division of Radiophysics of the CSIRO (Australia's largest national research organisation) and from the University of Sydney. The five-month survey covered the entire sky south of Declination +20, fully two-thirds of the celestial sphere. During the survey, the astronomers used the Mills Cross antenna to search for pulsar 'suspects'. The suspects were then observed using the 64-metre Parkes radio telescope operated by the CSIRO, to see if they really were pulsars. In all, 2500 suspects were detected from the Molonglo Observatory. However, many of them turned out to be multiple detections of 'bright' pulsars, or simple interference. Only 320 candidates were eventually observed from Parkes, resulting in a total of 155 new pulsars. The Molonglo team had doubled the number of known pulsars in one survey! The results not only allowed further study of individual pulsars, but also linked them to the distribution and frequency of supernovae.

By the early 1980s, pulsar astronomy was a hot topic, but it was about to get hotter. The detectors used to search for pulsars sampled the incoming signal about every 20 milliseconds. This meant that pulsars with periods less than about 100 milliseconds were difficult to detect. No one thought to go to any higher sampling rates because all the theoretical and observational evidence suggested that pulsars spin, at most, several times per second when they're first born. For example, the Crab pulsar was the fastest pulsar then known, with a period of 33.1 milliseconds. It was also the youngest, being the result of the famous supernova explosion observed by the Chinese a mere thousand years ago.

In 1982 Shrinivas Kulkarni was using the giant 305-metre Arecibo radio telescope to observe a radio source in the constellation of Vulpecula. The source had been discovered twenty years earlier, but later, more detailed maps of the area showed that the source was very small. This and other clues convinced a team of radio astronomers, led by Don Backer of the University of California, Berkeley, that it was a pulsar. In late October, Kulkarni confirmed the suspected pulsar, but this was no ordinary pulsar. With a period of 1.56 milliseconds, it was spinning at a phenomenal 642 revolutions per second—20 times faster than the Crab pulsar. The first, and to this day fastest, millisecond pulsar had been found.

The discovery of millisecond pulsars created even more problems for astronomers still trying to understand how 'normal' pulsars work. Here were objects spinning at up to 90 per cent of the speed needed to break them apart. Just how the neutron stars managed to get up to that speed was an even bigger problem.

When normal pulsars are first formed, they rotate much faster than we observe them today. They have magnetic fields one million times stronger than the strongest magnet on Earth. The high field is produced by the collapse process, which compresses and intensifies the magnetic field just as it compresses the matter. The strong magnetic field and powerful energy emission cause the spin rate of the pulsar to decline over a few million years. This helps astronomers estimate pulsar ages—the faster they're slowing down, the younger they must be.

Millisecond pulsars on the other hand, spin nearly a thousand times a second, which you might think indicates youth. Yet their slow-down rate is infinitesimal. This indicates that their magnetic fields are much weaker. In all respects they seemed quite different from those previously encountered.

So how do you make a millisecond pulsar? There

are a number of theories, but the most successful involves a binary star system. Imagine a binary system consisting of a massive star and a low-mass star. The massive star evolves faster, through the red giant stage, and finally becomes a supernova, leaving behind a neutron star. If the binary system remains intact, the system will consist of a normal star orbited by a neutron star, a pulsar.

Over time, the pulsar slows down as it loses rotational energy. As the period increases, the signal strength becomes weaker. After a few million years, it becomes invisible. By this time, however, the normal star will have evolved into a red giant. The expanded red giant star spills mass onto the slowly spinning neutron star. The matter spirals down onto the neutron star, causing it to increase in speed once more. It spins-up well past its original spin rate. Somehow during this process, its magnetic field becomes weakened. The new layers of matter may act like iron sheets wrapped around a bar magnet, which creates a very good magnetic shield. At the end of this process, the once dead pulsar has been reincarnated to become a millisecond pulsar. Only about one-tenth of a solar mass is needed to spin-up the pulsar to millisecond speeds. And because the magnetic field surrounding the millisecond pulsar is weak, it can last for billions of years.

This scenario is supported by a number of observations. For example, as mass is transferred onto the pulsar, the matter is accelerated to tremendous speeds. As the matter hits the surface of the pulsar, it can reach 10 million degrees, emitting X-rays as a result. This is just what's observed in X-ray binary systems like Scorpius X–1.

According to this scenario, all millisecond pulsars are formed in binary systems. Sometimes, however, the binary system consists of two high-mass stars. After the first pulsar has been formed and then spun up, the second star will also suffer a supernova explosion. This can disrupt the binary system, leaving one normal pulsar and one millisecond pulsar going their separate ways.

More than 700 pulsars have now been discovered, mostly in our little corner of the Milky Way. Almost all of them are quite good clocks giving very regular pulses, but slowing down steadily as their rotational energy is converted into high-energy particle beams and accompanying radiation. A few of them are spectacularly good clocks, rivalling the best clocks ever made here on Earth.

These precisely timed signals have allowed enormous advances in astronomy, such as the first detection of planets outside our solar system. Any change in the motion of a pulsar causes a Doppler shift in the pulse arrival times. If a pulsar is orbiting around another star, sometimes it is approaching us and sometimes receding from us, causing the pulses to appear more or less frequent, respectively. In a similar way, variations in a pulsar's period indicate the presence of objects orbiting around the pulsar, such as planets. Astronomers were astonished by the discovery of two planets, and then a third planet, around one single pulsar—could they have survived the supernova?

If the beam from a pulsar passes through the atmosphere of a stellar companion, the beam is delayed in a particular way that allows astronomers to study the nature of the companion star's atmosphere. In at least one case, the beam appears to pass across the star itself, causing it to be evaporated. The pulsar is blowing its companion away, heating a huge plume of evaporating gas like a comet's tail.

The space between us and the pulsars contains free electrons. This interstellar medium *disperses* the beams from pulsars—this means that high radio frequencies arrive before the lower frequencies. Further, the greater the distance, the more electrons the beam passes through and the greater the dispersion. By measuring the degree of dispersion, astronomers can estimate the distance to a pulsar based on a knowledge of the electron density between us and the pulsar.

The direction of some pulsars can also be determined to better accuracy than any other objects in space. The reason for this is that the Doppler shift of the pulsar signal also depends on the Earth's velocity around the Sun and its position relative to the pulsar. A direction error translates into an apparent annual or semi-annual variation in the pulsar period. If you noticed a variation with a period of exactly one year, you could pin it down to a pulsar planet with a period of exactly one year—very unlikely—or to a direction error—very likely. Once you know this variation, you can use it to refine the position of the pulsar in the sky to great accuracy.

In general, the accuracy of pulsar measurement is determined by whichever is worse: the pulsar clock, the terrestrial clock, or intrinsic fluctuations in the pulsar signal itself. Most pulsars have timing noise—fluctuations in their spin rate—sometimes due to huge, sudden changes in the rotation speed of the pulsar. These sudden changes, called glitches, are thought to be the result of a pulsar's equivalent of earthquakes. Pulsars have a thin solid crust. As the pulsar slows, its shape changes and the crust moves to accommodate the new shape. Variations in pulse rate may also be due to sudden change in the flow of the neutron fluid inside the pulsar.

For the most accurate pulsars, astronomers have to correct for the gravitational perturbations of both the Earth's orbit and the pulsar's motion. The motion of the Earth is perturbed not only by the Moon and the planets, but also by the asteroids. The combined effect of the 200 largest asteroids is detectable as a perturbation on the apparent rotation speeds of some pulsars. Other pulsars in globular clusters are easily perturbed by the gravitational field of nearby stars in the cluster. In both cases, the gravitational perturbation means that the relative velocity of the Earth and the pulsar varies erratically;

by the Doppler effect, this causes the apparent pulsar spin speed to change.

Pulsars may also offer an alternative way of estimating the rate of supernova explosions in the galaxy. As we've seen, young pulsars rotate faster than older ones; they also slow down faster. By observing how a pulsar behaves over time, it's possible to estimate its age. The trouble with pulsars is that the beams that make them so visible when they pass across the Earth can easily miss the Earth altogether. So it's important to estimate the 'beaming factor', that is how many pulsars are likely to be visible from any place in the Galaxy. The magnetic axis of pulsars is different from the rotation axis. This allows the radio beams to sweep the Galaxy, revealing the presence of the pulsar. Over time, however, the magnetic and rotation axes tend to align themselves, reducing the chances of detection. When it is aligned the beam no longer sweeps the sky, but just points steadily in one direction. The idea of pulsar alignment is easy to demonstrate. Take a hard-boiled egg and spin it with the long axis horizontal. The egg will spontaneously stand on end as the spin axis moves through the egg.

Along with this alignment process, pulsars also appear to get dimmer with age, eventually fading from view in about ten million years. The phrase 'appear to' is used deliberately, since it may be that a more important factor is that they simply move out of range, as we'll discuss in a moment. Combining all these effects, however, pulsars could be born as often as once every ten years in our Galaxy. Many astrophysicists are not convinced that pulsars do align, and find that they are born every thirty years. It's difficult to understand their scepticism, since all orbiting bodies change their spin axes with time due to asymmetries and gravitational forces from other bodies.

Another fascinating aspect of the pulsar population is their high velocities. The average speed of pulsars has been shown by the British radioastronomer Andrew Lyne to be about 450 kilometres per second. For comparison, the average speed of stars in the Galaxy is much slower, about 200 kilometres per second. This is the speed of the Sun, which moves around the Galaxy about once every 200 million years. The high average pulsar velocity means that most pulsars can escape the gravitational pull of the Galaxy, so it seems logical that most pulsars should be found in a huge halo around the Milky Way and other galaxies. They are born in the disk of the Milky Way, but get shot at a high velocity in a random direction. Thus most of them move out of the range of our telescopes over a few million years.

If most pulsars are created in supernova explosions, their high speed must mean that there was something asymmetrical about the supernova which gave each pulsar its kick-start. There are many possible explanations: for example, some phenomenon may cause neutrinos which are produced during neutron star formation to beam in one direction. Alternatively, one of the pulsar's two particle beams (one from each magnetic pole) could be stronger, creating a sort of rocket drive. Or the stellar collapse could be so asymmetrical that the neutron star is formed off-centre, blasting a much thicker surface layer from one side.

A likely explanation is that the high-speed pulsars are formed from the coalescence of two white dwarf stars. A pair of white dwarf stars, each about the size of the Earth, can orbit so close to each other that they become egg-shaped. Orbiting once every minute, their surfaces actually touch and material can flow from one to the other. As we've seen, white dwarfs are unstable: add mass and they collapse. In a binary system the star losing mass grows physically larger, so that its surface is closer to its companion, allowing even more matter to spill over to the denser companion. The star attracting the matter collapses even further, attracting still more matter from its bloated companion. All the while, the emission of gravitational waves causes the pair to spiral closer and closer. Sooner or later, the smaller, denser star takes the plunge and collapses into a neutron star. As it does so, it emits a blast of neutrinos that plough into the now-bloated companion star, evaporating the outer layers and sending the newly formed neutron star hurtling through space.

Only careful calculation or observation can distinguish between the various possibilities. If the last explanation is correct, you'd expect to find very high-velocity, low-mass white dwarfs flying through the galaxy. In all cases, there will be clear gravitational wave signatures which eventually should allow processes like these to be observed.

The enormous halo of neutron stars that we infer from the pulsar observations is similar in scale to the halos of dark matter that astronomers have been searching for. Based on the current rate of pulsar formation in the Galaxy—say once every thirty years—there simply cannot be enough of them to solve the dark matter problem. However, as mentioned earlier, many astronomers believe that in very young galaxies supernova explosions occur perhaps a hundred times more often than they do today. If so, the population of ancient pulsars might be much greater than we think, and they may be part of the solution to the dark matter mystery.

A few of the pulsars so far discovered consist of pairs of neutron stars orbiting each other with 8-hour to 12-hour periods. These binary pulsar systems provide the most exciting laboratory for general relativity. They satisfy all the requirements of Einstein's *gedanken* experiments: they have strong gravity, high velocity and are super-accurate clocks. They prove not only the accuracy of general relativity to a very high precision, but they have also proved the existence of gravitational waves. Only the direct observation of gravitational waves from the formation of black holes will provide a better testing ground for Einstein's theory. We will take up this story in the next chapter.

Chapter 8
Pulsars Prove Gravitational Waves

In the previous chapter we explored the rich diversity of pulsar astronomy. We saw that there is plenty of uncertainty: we are uncertain how they form, what drives them to high velocities, how many there are, and what gravitational wave signals they may produce.

But pulsars have already played a major role in the story of gravitational waves. In fact there is one binary pulsar which has been used to prove the existence of gravitational waves. The binary pulsar was discovered in 1974 by Joe Taylor and his then student Russell Hulse.

Taylor observed the system using the giant 300-metre-diameter radio telescope at Arecibo in Puerto Rico. The Arecibo telescope is nestled into a natural depression, the valley lined with metal reflectors to create an enormous dish. A huge gantry is suspended above the dish from cables attached to three tall pylons. The radio receiver itself is suspended at the focus of the dish. Although the dish cannot be moved, the receiver can be pointed to different parts of the stationary dish, allowing observation of different parts of the sky. The daily rotation of the Earth brings different objects into view. Although only a relatively small part of the sky can be seen using the Arecibo antenna, it can be seen with a sensitivity far greater than any other radio telescope on Earth.

Taylor and Hulse detected signals from the pulsar but not from its companion. By carefully examining the pulsar's period, however, they detected its companion star. Further study revealed the pulsar to be in a very eccentric orbit with the companion. During each orbit the pulsar would fall in towards the companion, and swing quickly around it before arcing back out into space. It repeats this slingshot orbit every $7\frac{3}{4}$ hours. The celestial dance is visible from Earth as a rhythmic change in the pulsar's period. When the pulsar is approaching, its pulses arrive more frequently; as it recedes, they are more widely spaced, a result of the Doppler effect. The pulse period—the time taken for the pulsar to complete one rotation—is shortest during the fastest approach and longer when it is receding. The pulse period is also altered by the fact that the radio signal has further to travel when the pulsar is furthest from us, and less distance to travel when it is nearest. The pulse arrival times vary by a few seconds due to this effect.

Fortunately, this pulsar has an extremely stable period. Unlike some pulsars, it has no glitches, no sudden changes in spin rate, so the pulse arrival times can be measured to an accuracy of a few microseconds. In its tiny orbit, the pulsar travels at some 300 kilometres per second—ten times faster than the Earth orbits around the Sun, and 0.1 per cent of the speed of light. These facts make it an ideal laboratory for studying general relativity: large masses, compact stars and relatively high velocities. What's more, one of the masses carries with it a super-stable clock which continually sends time signals to us. This is a relativist's dream come true!

For the binary pulsar, it is easy to calculate the expected changes in the pulse arrival times as a result of the various relativistic effects. Curved space should cause the pulsar's orbit to precess in the same way Mercury's orbit about the Sun precesses (Chapter 4). But instead of a few arc-seconds per year, the pulsar's orbit precesses about 4 degrees per year! That means that in about twenty years the eccentric orbit should have rotated almost 90 degrees.

Let's look at the pulse timing in more detail to see how the various relativistic phenomena can be studied. The pulsar system completes about 1000 orbits each year, or 20 000 orbits in twenty years. In the absence of curved space, you would expect the orbital pattern to be stationary in space. Imagine that we are looking at the pulsar system from end on, along the axis of the orbit. When the pulsar is falling towards its companion, it moves faster and faster; each pulse is emitted towards us when the pulsar was a bit closer to us than the last, so that the pulses appear to be coming more frequently. Just one cycle later, this would be repeated exactly the same as the first, and if space were flat the pattern would be repeated indefinitely. But because space is curved around the pulsar, the orbits don't quite add up and the orbital pattern rotates. About 20 000 cycles later, the pulsar orbit is side-on to us. Now, as the pulsar falls towards its companion, the pulse frequency is close to its average value, with a brief change during the swing-by when it is moving either towards or away from us (depending from which side we're looking). In this way it is relatively simple to monitor the way the pulses vary and learn about the relative orientation of the pulsar system, and so observe the precession.

At times in its orbit, the radio beam from the pulsar passes closer to its companion star than at others. The companion star has a similar mass to the pulsar. When the pulsar beam passes the companion, it must follow the local curvature of space-time. The large deflection means there is a delay in the arrival of the signal. This small delay, due to the changing sampling of the companion's local space-time curvature, adds to the delay due to the orbital motion. Although the radio beam takes the shortest possible route, it is still longer than if there were no deflection of space-time due to the companion.

Another effect arises because of the eccentricity of the pulsar's orbit. When the pulsar is close to the companion, its pulsed beam has to climb out of the gravitational well created by both stars. At these times the gravitational redshift is large compared with periods when the two stars are farther apart: that is, the pulses arrive less frequently.

The binary pulsar has allowed all these tiny effects to be measured with great precision—they agree precisely with general relativity—and in the process other information can be gathered. For example, the masses of the stars can be measured. One has a mass of 1.441 solar masses, while the other has a mass of 1.387 solar masses. These measurements represent the most accurate mass determination of any object outside the solar system. It is amazing that the small, subtle effects of general relativity allow us to measure the grossest and most fundamental properties of these stars to such an accuracy. It is fascinating, too, that these masses are so close to the Chandrasekar critical mass limit for white dwarf stars. Does this mean that they formed from the collapse of white dwarf stars, and not from the supernova collapse of very massive ones?

The binary pulsar, like any other binary system, must give off gravitational waves. Because of the eccentricity of the pulsar orbit, these are mainly in bursts lasting half an hour or so every $7\frac{3}{4}$ hours, as the two stars swing past each other at closest approach. The gravitational wave energy emitted by an object rises dramatically with orbital speed. So, in the binary pulsar's eccentric orbit, the gravitational wave energy is vastly greater during the close approach than at any other time in its orbit. In this system, there should be bursts of gravitational waves of about one-thousandth of a cycle per second, lasting for half an hour each orbit.

Detection of these gravitational waves from the Earth

would be very difficult; our detectors are not sensitive enough. However, the system should lose a substantial amount of energy in gravitational radiation. Gravitational waves should extract energy from the system in much the same way that viscosity of water extracts energy from you when you're swimming. As the system loses energy, the two neutron stars should spiral closer together, the orbital period shrinking as they do so. The change in velocity caused by the emission of gravitational waves can be easily calculated: the $7\frac{3}{4}$-hour orbit should be shortening by 70 microseconds per year. Joe Taylor and his collaborators have been measuring the orbital period of the binary pulsar for about twenty years and have confirmed this prediction. The binary pulsar is, indeed, emitting gravitational waves! It was fitting that Taylor and Hulse received the Nobel Prize for their work in 1993.

As time passes, the binary pulsar will orbit faster and faster. As it speeds up, it will emit stronger and stronger gravitational waves, causing it to lose energy at a faster rate and hence speed up even more. Eventually, the two neutron stars will coalesce. This will happen in about 300 million years. Three years before they finally merge, their orbit will have contracted so they are orbiting every three seconds. By this stage, the two stars will be a mere 1000 kilometres apart, still many diameters away from each other, but the whole system would easily fit inside the diameter of the Earth. The two stars will be tearing around each other furiously at more than 1000 kilometres per second.

A minute before coalescence the stars will be whipping around each other 15 times a second, but the final merger will occur only in the last few milliseconds, when the stars are spinning around each other hundreds of times each second.

Here we have an immensely powerful and distinctive source of gravitational waves. It is distinctive because it is a 'chirrup': a note rising steadily in frequency and amplitude. It can be accurately predicted throughout the coalescence stage. At its last moments, it becomes extremely intense. For example, when the stars are spinning about each other 500 times a second, they are only 30 kilometres apart and travelling at one-sixth of the speed of light.

Just before coalescence, the binary pulsar emits gravitational waves with a 'luminosity' of a hundred thousand galaxies! As they merge, the gravitational wave luminosity continues to increase. Just how bright these gravitational waves become depends on the physical properties of neutron stars: will they tear each other apart, or remain intact to the bitter end? The answer is unknown.

The last few milliseconds of the gravitational wave signal carries details of the neutron star matter: its viscosity, compressibility and density, and perhaps information of the quarks and strange particles that are present along with the neutrons. The gravitational wave signal contains details of the fragmentation and re-coalescence, the disk or bar modes which form, and the way the neutron stars oscillate.

Finally, since the total mass of the coalescing binary exceeds the neutron star critical mass, as most people suppose, we should be able to follow the system as it collapses to form a black hole. This in itself should have a distinctive signature, constituting the only clear way we are ever likely to have to image a black hole.

Whether or not we can really observe neutron star coalescences depends on how often they occur, and how close they are. Right now there are only three known binary neutron stars that are likely to coalesce within the present age of the Universe. One, PSR 1913 + 16, is roughly 30 000 lightyears away; another, PSR 1254, is only 3000 lightyears away; the third is in a rather distant globular cluster.

Many more binary pulsar systems have been found in longer, slower orbits (more than a few days). The time before they coalesce is longer than the age of the Universe so they are not interesting candidates as sources of gravitational waves. Other fast binaries consist of neutron stars orbiting normal stars or white dwarfs, like the system where the pulsar beam is blasting its companion away. In many other cases, matter from the normal star is being 'sucked' onto the neutron star by the pulsar's gravity. As the matter spirals down onto the neutron star like water down the plug-hole in a bathtub, the matter heats up to tremendous temperatures and emits X-rays. In all these cases, the dynamics of the system are dominated by the mass exchange, obliterating the effects of general relativity and gravitational waves.

However, we can be quite certain that the three relativistic neutron star binary systems that have been found are only the tip of the iceberg. Firstly, pulsars have very narrow beams, so you have to be lucky enough to be in the line of sight in order to see one. Secondly, pulsars don't shine forever, and slowly fade from view. Thirdly, binary pulsars are even harder to find because the pulse period is not constant, but varies as they orbit. (It is much harder to search for pulses of varying frequency, especially if the pulses are faint.)

No one knows how often binary pulsars merge in the Galaxy, but if they live as a pair for 300 million years—twenty times less than the age of the Galaxy—we can be confident that throughout the Universe there are many such binary pulsars at all stages of evolution. There are probably 100 000 binary neutron star systems in our galaxy alone, and there is a moderate chance that one of them is within about 10 000 years of coalescing. A star system at this stage would be orbiting once every few minutes and the stars would be moving at about 0.5 per cent of the speed of light. So, what we see now are not all the pulsars that exist in the Galaxy, just those few that are visible. There could, in fact, be between 100 and 1000 binary pulsars like the Hulse–Taylor pulsar actively emitting radio beams in the Galaxy. There could be as few as 1000 or as many as 100 000 of them, mostly no longer beaming (because the pulsar has switched off or aligned itself). All of these would be on their way towards coalescence. But even 100 000 pulsars is not enough if the average time to coalescence is 300 million years. What we need is a sample of 300 million binary pulsars. Then, on average, one should coalesce each year.

If we had detectors capable of detecting gravitational waves from a binary coalescence within 150 million lightyears, we could be pretty certain of seeing more than one per year, and this must be the goal for any detector which we want to be certain of detecting gravitational waves. Unfortunately, a gravitational wave arriving from such an event would be tiny. The quadrupole deformation it would create in detectors would be like trying to measure a change in the distance from the Earth to the Sun of one-tenth the diameter of an atom. While your intuition may say this is impossible, we will see that it is possible to detect such feeble signals.

As well as producing gravitational waves, pulsars may also be useful in detecting gravitational waves produced by other events far across the Universe. We've seen how gravitational waves make masses move in a characteristic quadrupole deformation. Now think about this situation: a pulsar is sending out an extremely regular timing signal. While the pulses vary due to orbital motion, this is very predictable. What isn't predictable is the change in pulse arrival times caused by a change in distance between Earth and the pulsar, which can happen if a gravitational wave passes by.

If a burst of gravitational waves were emitted somewhere in the Universe with a frequency of one cycle per second, the waves would spread out like ripples in a pool, perhaps reaching the pulsar first and, depending on the distance, reaching the Earth a thousand years later. What will the gravitational waves do to the pulsar signal?

The waves will have short wavelength compared with the Earth–pulsar distance, and it makes little sense to worry about the change in distance between the two. It is better to think about the interaction between the gravitational wave and the pulsar radio signal. The gravitational wave is time varying curvature, so the radio path will be a little bent by the gravitational wave. This will cause the radio wave to be successively advanced and delayed as the wave passes. But because the radio signal path is so long, the advanced and delayed sections will all add up to nothing. The effects cancel out all along the path from the pulsar to the Earth. But not quite all. During the last half wavelength of the gravitational wave, the effect does not cancel out, and the radio waves from the pulsar will be affected. By comparing the pulse arrival times with a very accurate clock on Earth, you could detect the gravitational wave.

If a few more pulsar signals were also being monitored, the gravitational wave would affect each pulsar in turn as it moved across the Galaxy. Correlating the data from several pulsars would not only reveal the presence of the gravitational wave, but also the direction of its source. We would have a gravitational wave telescope. Searches of this sort are presently being conducted, but so far nothing has been found. Pulsars are best for detecting very low frequency gravitational waves, gravitational waves almost as slow as the tides in the ocean, say one cycle per day, which could be produced when billion-solar-mass black holes collide, as they might do when quasars collide in the early Universe.

CHAPTER 9
BLACK HOLES AND THE BEGINNING OF TIME

Pulsars are the result of the collapse of matter into an incredibly small space. The mass of an entire sun is squashed into a ball the diameter of a city. Bizarre as such an object might seem, nature has gone one better: black holes. If even more matter is added to a neutron star, there is a point where the neutron star's increasing gravity overcomes the degeneracy pressure of the neutrons, allowing the star to collapse still further. The end result is an object of infinite density called a singularity. Surrounding the singularity is a black hole.

To understand what a black hole is, try jumping off the Earth! Because the Earth has mass, you need to attain a minimum speed to escape from Earth's gravity. Rockets must reach this 'escape velocity' in order to fly to the Moon, for example. The escape velocity on more massive planets is correspondingly higher. Now, imagine an object so massive that the escape velocity is greater than the speed of light. This is just what happens around a singularity and nothing, not even light, can escape from this black hole in space.

The idea of a star so dense that not even light can escape from it is not new. They were first considered by the British natural philosopher John Mitchell in 1783: he called them 'dark stars'. Mitchell calculated their size and imagined a Universe with a sizeable population of such invisible objects. Our present name for them, black holes, was first coined by John Archibald Wheeler in 1967.

Einstein's equations had predicted black holes, then called 'Schwartzchild singularities', named after Karl Schwartzchild (1873–1916), a talented German astronomer. Einstein disliked the idea of a black hole so much that he was determined to prove they were impossible. In 1939, Einstein published a paper in the *Annals of Mathematics*, in which he concluded that 'Schwartzchild singularities do not exist in physical reality'. But he was wrong. His prejudice blinded him to the possibility of the existence of black holes. He assumed that gravitational collapse was not possible, and yet this is precisely the mechanism that allows black holes to form.

Electron degeneracy pressure supports white dwarfs up to a mass of 1.44 solar masses. Beyond this mass a white dwarf collapses to form a neutron star, which is supported by neutron degeneracy pressure. The enormous gravity of a neutron star means that its escape velocity is about 10 per cent the speed of light.

But what happens if you add even more mass? In the same year that Einstein published his flawed proof of the prohibition of black holes, Robert Oppenheimer and his student Hartland Snyder of the University of California at Berkeley published a study of the gravitational collapse of an idealised spherical star. They created a mathematical model of the star's collapse and uncovered the extraordinary contrast between what is observed by an observer riding in on the collapsing star's surface and what is seen by a distant observer. The observer riding in on the collapse watches the star shrinking faster and faster, unstoppable. But the far-off observer sees the shrinking process rapidly grinding to a halt as the gravitational redshift stretches out the time intervals until its final observable moments are freeze-framed in infinitely redshifted radiation. Never before had the effects of relativity and of different reference frames been so dramatically illustrated.

Oppenheimer and Snyder's calculations were distrusted for many years because of the idealisations of their model. It took the physicists of the US hydrogen bomb programme, led by Sterling Colgate, to provide the methods and the computers for realistically simulating gravitational collapse. In 1963, they provided the first conclusive proof that gravitational collapse could create black holes. This made it clear that black holes are not prohibited, and very soon astronomy provided the first tentative evidence that black holes do 'exist in physical reality'.

Black holes form because there is no force or pressure that can stop a neutron star from collapsing once it exceeds its critical mass. Just what this critical mass is remains uncertain, but it is definitely less than 3 solar masses. Once a neutron star has collapsed to just one-tenth of its original size, its escape velocity is greater than the speed of light. If light cannot escape, nothing else can, and there is no way an observer on the inside can tell us anything else about it.

If you sent a flashing torch down into a black hole, you would see the flashes repeating more and more slowly, and the light becoming redder and redder as a result of the gravitational redshift. Quite suddenly, the time between flashes will be stretched to infinity. In other words, the flashing ceases; the torch disappears from this Universe forever. This discovery was made by Snyder and Oppenheimer, and the point at which the torch disappears is called the event horizon. The event horizon is, therefore, a spherical shell around the singularity beyond which nothing is visible . . . or recoverable.

The size of the event horizon is called the Schwartzchild radius, named after Karl Schwartzchild. Within weeks of their publication in 1916, Schwartzchild applied Einstein's equations to a non-rotating spherical star. On war service for Germany against Russia at the time, he found his solution, the Schwartzchild metric, and immediately posted it to Einstein who presented it at the Prussian Academy of Sciences in January of the following year. By June, Schwartzchild was dead, but his name lives on in every black hole.

If you went down to investigate what had happened to your torch, you wouldn't find any particular barrier at the Schwartzchild radius. While we have clues as to what happens beyond, there is no way of ever knowing for sure. If you went beyond the event horizon and did see what lies there, you'd know, but you could never get word to the rest of us outside. More gruesomely, you'd never return either.

In 1965, Roger Penrose (then at the Birkbeck College, London University) published a proof that black holes must contain a singularity, a point where the density of matter reaches infinity, where gravitational tidal forces are also near infinity, and where the gravitational theory of general relativity is replaced by quantum gravity. At the singularity, which is about 10^{-35} metres across, space and time lose their meaning. All of the matter in the black hole is concentrated in this singularity, the rest is empty space.

Is it possible to enter a black hole and survive to tell the tale? In the case of a black hole as heavy as a typical star, the answer is no. If you were falling towards a black hole, the gravitational pull on your feet would be greater than on your head. Nasty as it sounds, your feet would be pulled off your legs, followed by the rest of you in quick succession. The same fate awaits any scientific instrument you drop into a black hole. However, when black holes become big enough, like the billion-solar-mass black holes suspected to lie at the hearts of many galaxies and quasars, at least one-way journeys do become a possibility.

When matter falls into a black hole, all information

about its composition is lost. You could pour in iron, light, helium or neutrinos and the end result would be the same: the black hole would get heavier, but you'd never know what it was made of. This was proved by Brandon Carter, Werner Israel and Stephen Hawking in the famous 'no hair' theorems which state that 'black holes have no hair'. This means black holes are devoid of any external details, bald if you like. The only things you can ever learn about a black hole are its mass, its angular momentum, and its electrical charge.

Black holes can vibrate, however. Kip Thorne, in his book *Black Holes: The Membrane Paradigm* (based on work by Thibault Damour and others),[1] shows that you can think of a black hole as a bit like a balloon covered in a thin membrane. Chadrasekhar and Detweiler have calculated all the vibration frequencies of black holes. If an object falls into a black hole, it sets up a vibration like a stone hitting a balloon. The vibration is very strongly damped by gravitational wave emission, except in the case of very rapidly rotating black holes.

Now think of the vibration in the context of a central singularity. How can something, empty except for a single point, sustain vibrations all the way to the surface? The answer comes back to space: the vibrations are vibrations of space, not of the matter inside it. Much of the energy of a black hole is that of the curved space-time that makes it and surrounds it. The vibrations of black holes are the vibrations of this space-time, not of the singularity at its centre.

If you approach a black hole in a spaceship, you discover some strange effects. For a start, you find yourself being dragged around in the same direction as the rotation of the black hole. This is the same frame-dragging effect we talked about in Chapter 4: while it's barely detectable in our solar system, it dominates the dynamics near a rotating black hole. No matter how much you struggle against the pull with your rocket engines, you find yourself carried irrevocably in the direction of the black hole's rotation. Your movement through space is as irresistible as your journey through time. Simultaneously, however, your efforts to escape do not result in any change of trajectory, but rather in a variation of time intervals. You find yourself moving about in time (always forward, but at different speeds) but not in space. This is gravitational time dilation enormously amplified. For you, time has become space-like, while space has become time-like.

As matter falls into a black hole, it is compressed tremendously: it becomes hotter due to friction, and it emits X-rays and possibly triggers nuclear detonations. In our galaxy, there are many systems where matter from a normal star is flowing over into a black hole. The approaching matter spirals around the black hole in an accretion disk. An accretion disk looks a bit like Saturn's rings, but instead of being very thin and circular, it is thickened and the matter spirals down to the surface. It is a three-dimensional version of what happens when water flows down a bath's drain-hole. There are probably many thousands of black hole systems in our galaxy, but often the flow of matter stops and they disappear from view. In 1994 two such systems suddenly flared up, producing powerful bursts of gamma rays as well as jets of material that could be seen spurting out at close to the speed of light. In other systems, the X-ray emission fluctuates violently, perhaps because the accretion disk oscillates as it is tilted this way and that during its last tight orbits around the black hole before finally plunging down into oblivion.

There has been much speculation about whether black holes can be used to create *closed time-like loops*. Closed space-loops are simple: run around in a circle and you have created one. A closed time-like loop would allow you to travel both forwards and backwards in time. In other words, you would have a time machine. Some people (such as Oxford physicist David Deutsch) claim that the laws of physics do not prohibit such loops. However, behind and supporting the laws of physics are great meta-laws: laws of logic, causality and self-consistency. The Universe must be logical, causal and self-consistent. In our everyday lives, we all act in this belief. Even the act of switching on a light embodies our belief in these laws. The meta-laws ensure that the normal laws of physics do not contradict each other, nor vary from place to place in the Universe. Stephen Hawking's explanation is very good. He says that time machines are prohibited by the Universe's 'chronology protection agency'! The purpose of the CPA is 'to keep the Universe safe for historians'. And, says Hawking, we have excellent experimental proof of this by the fact that we are not deluged with hordes of time tourists from the future! With this in mind, we should look very carefully at the laws which suggest time-like loops—they must be wrong because they violate the meta-laws.

Another speculation about black holes is called the cosmic censorship conjecture, proposed by Oxford mathematician Roger Penrose. As Penrose proved, black holes have singularities at their centres where the density becomes infinite. Penrose asked: is it possible to have a naked singularity, one which is not surrounded by an event horizon. Or does the Universe always prudishly protect a singularity by clothing it in an event horizon so that it can never be seen?

One way a naked singularity could form would be if a black hole evaporated. The idea of evaporating black holes was discovered by Stephen Hawking in about 1974. It rests on the idea of vacuum fluctuations, those random, unpredictable oscillations that always exist, always fill space, and are ultimately responsible for quantum uncertainty. When a black hole is small, there is a significant possibility of a vacuum fluctuation creating a pair of *virtual* particles near the event horizon. Virtual photons and virtual gravitons are particle-like manifestations of vacuum fluctuations. They can momentarily appear and disappear, as long as their appearance is so brief that their energy cost is not recognised by the rest of the Universe that polices the law of conservation of energy! That is why they are said to be virtual.

But near a black hole, any pair of virtual photons experiences the strong gravity gradient of the hole. This tends to pull one photon towards the black hole faster than the other, so the pair is torn apart; in just the same way that a black hole can tear your feet from your legs as you fall inward. Once pulled apart, the virtual photons become real photons. For each pair of photons, one can escape the hole; the other one falls into the hole, returning half the acquired energy. The other half is paid for by the hole, by a reduction in mass. In this way, Hawking showed, black holes are surrounded by radiation and are slowly losing mass. The smaller they are, the greater the probability of a pair of virtual photons or gravitons appearing in the right place, and the greater the radiation. As they lose mass they get hotter, eventually exploding with the energy of a billion megatons of TNT. The reason we don't see black holes exploding all over the sky is that they have to evaporate down to atomic dimensions before the cycle of evaporation, shrinking, faster evaporation and faster shrinking can cause them to explode in less than the age of the Universe. If black holes with a mass of 10^{11} kilograms were formed during the Big Bang, we might expect to see them exploding today. So far there is no evidence of this happening.

If a black hole evaporates, what happens to its singularity? No one really knows. Some think it could remain behind, exposed to the prying eyes of the Universe; others believe it must cease to exist.

Now that we have a fair understanding of the nature of black holes, let's turn to the question of whether black holes can be sources of gravitational waves. In particular, can a coalescing pair of black holes produce gravitational waves? Because of gravitational wave emission, all orbiting systems must slowly spiral together. For systems like our solar system, this process will take billions of times the age of the Universe. In the case of close-spaced black holes, the whole merger takes place rather more quickly and, at least in terms of gravitational waves, more brilliantly than most other phenomena in the Universe.

Imagine two black holes in orbit, slowly spiralling towards one another. Since the escape velocity of a black hole is the speed of light, the two will meet at this tremendous speed. When they merge, the gravitational wave power produced is equal to the fantastic figure of 10^{52} watts. This figure is often referred to as 'the luminosity of the Universe', since it is roughly equal to the total power emitted by every star in every galaxy in the Universe. This energy level can't be sustained for long, however, and in the case of the merging black holes it lasts only a millisecond.

You might imagine that, if two black holes spiral together, then all the available gravitational energy—roughly the total mass energy of one of them—could be converted to gravitational waves. In practice, you don't get such 100-per-cent efficiency, because once a gravitational wave has been created it then has to escape the influence of the black hole. Gravitational waves are gravitationally redshifted, just as light would be. The waves produced very close to the black hole are extremely redshifted, while those produced further away have less redshift. All told, more than half of the gravitational wave energy is lost in this way, while more ends up inside the black hole, adding to its mass. The highest possible efficiency for converting mass energy to gravitational wave emission is thought to be just over 40 per cent. The intense gravitational waves can only be emitted in very short bursts.

What about the frequency of the gravitational waves? In the case of the normal binary stars, so little energy is produced that the systems are able to sustain their orbital motions for extremely long times. This means that the gravitational wave frequency is relatively steady at twice the orbital frequency of the stars—say, one cycle per day. For near-contact white dwarfs, the frequency could be as high as one cycle per minute, before coalescence. Neutron stars are able to sustain orbital motion up to almost 500 cycles per second. At this speed, the gravitational wave power, at a pitch about two octaves above middle-C, becomes very large and the system is extremely short-lived: the stars coalesce rapidly. For a pair of merging solar-mass black holes, the frequency could rise to more than one thousand cycles per second, but this will be sustained only momentarily at maximum power. The wave in this case steadily rises in frequency, getting stronger and stronger, rising to an extremely sharp, short spike as it approaches the luminosity of the Universe, followed by a brief ringing of the oscillating black hole. This ringing signal allows the new black hole to be imaged in the momentary shimmering of its birth.

While the merging of solar-mass black holes produces very intense, high-frequency gravitational waves, the coalescence of the enormous black holes thought to exist in the centres of some galaxies produces low-frequency gravitational waves. This does not happen often in our corner of the Universe, but photos taken by the Hubble Space Telescope show very young galaxies in the throes of violent collisions. Such processes may often involve the collision of huge black holes that would show up best as long, slow gravitational wave pulses.

There is one final source of gravitational waves that we should look at briefly. It is the most exciting; it is also the most speculative: gravitational waves from the origin of the Universe.

Many cosmologists believe that, in an infinitesimally brief period during its birth, the Universe expanded in a process called inflation. By this theory, the observable universe is a tiny cell in an enormous mega-universe. One consequence of inflation is that any gravitational waves present can be amplified by the process. Since the Big Bang is thought to have emerged from a fluctuation in space-time, we expect plenty of gravitational waves to have been present. If so, then today we should see space-time filled with gravitational waves. If these could be detected, they would carry the signature of the Big Bang when the Universe was less than one-billionth of a billionth of a second old. Unfortunately, the size of these ripples is quite uncertain. They could be more or less easily detected, or they could be beyond all current technological capability.

For the last forty years, high-energy particle accelerators have been used to probe smaller and smaller subatomic structures, using particle beams of higher and higher energy. Today these machines have exposed fundamental structures down to the level of quarks, which come together in threesomes to make protons and neutrons. The energies achieved in particle physics laboratories on Earth existed during the first nanoseconds of the Big Bang. The discoveries of particle physics have therefore allowed theoretical study of the early Universe back to its first microseconds and nanoseconds. These studies go on to make predictions about the Universe as a whole, which can be compared with observation. The most stunning example of this is the prediction of primordial gas concentrations in the Universe. Theory predicts 76 per cent hydrogen and 24 per cent helium, just the concentrations observed by astronomers in the primordial gas which has not yet been processed in stars.

Unfortunately, particle physics seems to be getting close to the limit of achievable energy because nobody is likely to pay for the construction of machines as large as the Earth, or the solar system. This means that even higher energies, closer to the moment of creation, are also out of reach.

Here, cosmological gravitational waves offer a thread of hope. They would originate not in the first microsecond of the Universe but in the first one-billionth of a nanosecond, when the particle energies of the cosmic fireball were a billion times higher than any laboratory could create on Earth. Thus the faint whispers of cosmological gravitational waves represent a probe of physical laws at energies billions of times higher than can ever be directly achieved. This may include the information needed to understand how and why the Universe began.

We have now explored the major sources of gravitational waves in the present Universe; that is, we have looked at those which physicists are most confident in predicting. Supernovae are likely sources of gravitational waves, as are binary pulsars. We have seen that other, weaker signals from distant supernovae and from the Big Bang create a continuous background of ripples in space-time. From this we can be certain that space-time not only curves predictably in response to matter, but that all of this curvature is impressed on a space-time which is randomly and unpredictably fluctuating from events unimaginably distant from us in space and time. The strategy to detect and monitor these unpredictable fluctuations has kept a growing community of physicists busy for a quarter of a century. It's now time to meet these physicists, and take a look inside their silent laboratories as they listen intently for the vibrations of space.

Notes

1 *Black Holes: the Membrane Paradigm* edited by K. S. Thorne, R. H. Price and D. A. Macdonald, Yale University Press, London, 1986.

Chapter 10
The Searchers

In 1969, an American physicist, Joseph Weber, announced he had achieved what many had presumed impossible: the detection of gravitational waves. The announcement propelled him into the limelight, and Weber was soon in great demand to give lectures all over America. Thin, athletic, with strong features and a head of wiry, greying hair in the Einsteinian tradition, he gave lectures which were energetic and full of jokes. Weber even made mathematics funny, so that students often left his lecture with their sides aching from laughter. They also left inspired.

But Weber's fame would soon be tainted with doubt. Although he continued to assert that his discovery of gravitational waves was real, many began to doubt the validity of his results. The turmoil that followed is a case study in the terrestrial nature of science, even the branches that search for the answers to cosmic riddles. Whether or not Weber actually discovered gravitational waves is something we'll talk about shortly. One thing can't be denied, however: this great scientist inspired the world to search for one of the remaining unconfirmed predictions of general relativity.

After World War II, physicists and engineers began exploring a new way of amplifying electromagnetic signals. The most exciting discovery to come out of this research was the transistor, that tiny, electronic silicon switch that has powered the electronic revolution ever since. Right in the thick of this research was the young Joseph Weber. Fresh out of the navy, Weber looked for ways of amplifying signals using systems 'which were not in thermal equilibrium'. His idea was to use collections of atoms or molecules that were energised in such a way that when a weak radio signal passed through them, the atoms emitted energy in exactly the same way as the original signal, only stronger. The physics behind the process was first predicted by Einstein and eventually earned the acronym MASER, for Microwave Amplification by Stimulated Emission of Radiation. Better known is the visible-light equivalent called LASER. The discovery earned the Nobel Prize for Charles Townes, Nicolay Basov (1964), and Aleksandr Prokhorov (1964). While Weber most certainly was in the running for a share in the award, he was left out, and very disappointed. At the end of the 1950s, Weber switched fields. Quite alone, he began a new line of research: the detection of gravitational waves. While most scientists, including Einstein, had dismissed gravitational waves as being of merely academic interest—how could such faint signals possibly be detected?—Weber took on the challenge.

Gravity interacts with mass, so a good gravitational wave detector is a big lump of matter. Since gravitational waves have a characteristic quadrupole effect, the matter could be of almost any shape: a cylinder, a sphere, even a bell can be stretched and compressed in a quadrupole motion. But whatever the shape, it must be big. Weber decided to use a cylinder, or bar. When a gravitational wave passes through a metal bar such as the one Weber envisioned, it should preferentially excite the long axis of the bar. In other words, the two ends of the bar would start moving in and out like two masses connected by a spring. This is why the bars are called resonant bar antennas: they vibrate in sympathy with the passing gravitational wave. Further, depending on the size of the bar, it will preferentially vibrate at a certain frequency.

No matter how large the test mass, however, the problem of how to detect the tiny vibrations caused by a passing gravitational wave remained. Weber approached the problem using the properties of piezoelectric crystals. These fascinating materials have the property that when they're deformed they develop a large voltage. Such materials are used in such widely varying applications as gas lighters, watches, earphones and beepers. These materials are amazingly sensitive to vibrations, almost perfect for what Weber wanted. Attached to a test mass, they would signal the passage of a gravitational wave by detecting the deformation in the mass as the wave passed. The deformation would cause the crystal to change shape, emitting electrical current as it did so. The current could then be observed and recorded, allowing the observer to study the nature of the gravitational wave.

Depending on their size and shape, piezoelectric crystals can produce perhaps a million volts for every millimetre deflection. This is equivalent to 1000 million volts per metre, so if you attached piezoelectric crystals to a big metal bar, say a metre long, a one-volt signal might represent a vibration on the order of one-ten-thousandth of a millimetre.

But in every electrical system there's something called 'noise'. Because the amplitude of the signals likely to be produced during the passage of a gravitational wave is so infinitesimal, noise always presents a problem for detecting gravitational waves. Fortunately, Weber's work on amplifiers prepared him to address the problem of detecting very small signals. To give you an idea of what noise is, outside its usual meaning, think about your stereo system. Your hi-fi system is designed to detect and convert into sound waves a range of vibrations from 20 hertz[1] (20 vibrations a second—the lowest we can hear) to almost 20 000 hertz (the highest pitch young people can hear). This is a total range—or bandwidth—of 19 980 hertz, or close enough to 20 000 hertz. If you turn on your stereo and turn up the volume as high as possible (without the radio, turntable or CD-player running!), you'll hear a characteristic hiss. This is what electrical engineers call 'Nyquist noise'. It is often called 'white noise' because it sounds the same at all frequencies, analogous to white light which is a combination of all visible electromagnetic frequencies. Nyquist noise in your stereo is created by the random vibrations of electrons within the components of the system. The amount of noise depends on the resistance of the components, their temperature, and how often you sample the signal.

There's one way of reducing noise in a system such as a stereo: reduce the bandwidth. You can prove this to yourself by turning down the tone control and so cutting out the high frequencies. This removes a lot of the hiss, but at the expense of the crispness and clarity of the music. If you reduced the bandwidth to 1 hertz (impossible with a stereo or radio) you'd reduce the noise down to perhaps 10 nanovolts, or ten-billionths of a volt, compared with the usual noise level of about one microvolt (100 times larger). But now the signal would be no more than a constant tone of varying loudness. You could choose where in the spectrum your tone was located—it could be a low hum or a high whistle—but at best the only information you could send would be something like Morse code. You certainly couldn't transmit speech or music. This emphasises a fundamental dilemma in physics: increase the bandwidth to get more information and you increase the noise!

Weber was faced with this dilemma when trying to devise ways of measuring the vibration in a big piece of metal. Thanks to the spectacular sensitivity of piezoelectric crystals, it turned out to be not much of a problem.

The noise level of 10 nanovolts described above translates to a minimum detectable strain in a 1-metre bar, or less than the size of the nucleus of an atom. That means, if you are willing to do a crude morse-code type measurement, you can detect vibrations as small as an atomic nucleus. Not a bad start towards building a gravitational wave detector.

Yet all solid objects vibrate, a result of the random motions of the atoms of which they're made. In a metre-long bar, the level of vibration is about four times larger than the smallest vibrations detectable by the piezoelectric crystals. To get around this problem, Weber decided to use a bar of aluminium which had a very high 'quality factor'. Quality factor is a measure of a resonator's ability to ring for a long time. A good example of objects with high quality factors are bells and tuning forks. Objects with low quality factors, like squash balls, convert vibration energy into heat. This is why a squash ball is warm at the end of a game. The reason a bar with a high quality factor would help is that while it would be constantly vibrating, these vibrations would be like a steady hum, their amplitude varying slowly with time. Only when a gravitational wave passes through would there be a sudden change in the amplitude of the vibrations, signifying the event.

There is yet another source of vibrations that Weber had to allow for—vibrations from the environment. In extreme cases this means earthquakes, but far more commonly it means vehicles, slamming doors and all the noise of human activity. Weber reasoned that since seismic vibrations travel at a few kilometres a second, two detectors placed thousands of kilometres apart would never feel a seismic disturbance simultaneously unless it happened to be exactly halfway between them. Gravitational waves travelling at the speed of light, on the other hand, would arrive at each detector almost simultaneously.

And so Weber built two gravitational wave detectors. Each consisted of a large aluminium bar weighing several tonnes, with piezoelectric sensors attached to the outside. They were mounted in a vacuum chamber and isolated as well as possible from external vibrations. One detector was in a bunker on the golf course at the University of Maryland. Its length gave it a frequency of a bit over 1000 hertz, like a high note on a piano, and about the frequency of gravitational waves expected from the formation of a black hole of a few solar masses. The second was built at the Argonne National Laboratory in Chicago. Being a thousand kilometres away from each other, they were perfectly placed to either confirm or discount the detection of a gravitational wave by either one. Weber set up a telephone connection and recorded long charts showing the vibrations of both detectors. At first searching the charts visually, and later with the aid of computers, Weber waited for the passage of a gravitational wave.

As time passed, Weber realised that a number of accidental coincidences showed up in the strip chart recordings. These occurred due to the random variation in the ringing of the bars and all the various noise sources. To overcome this problem, Weber displaced one of his charts by up to a minute compared with the other. An analysis on the two charts so far out of synchronisation should reveal only truly random coincidences. By comparing these 'time-delayed coincidences' which could not possibly be gravitational waves, with zero-time-delay coincidences which might contain gravitational wave signals, an excess of zero-time-delay signals should show up. These would be evidence of gravitational waves.

In late 1969 Weber published his first data in the prestigious *Physical Review Letters*. There in his data were an excess of zero-time-delay coincidences, the signature of true gravitational waves. The result caused enormous excitement. The relatively poor sensitivity of his two detectors meant that every pulse he detected represented flashes of gravitational waves millions of times more powerful than anything anyone had expected. Theorists imagined that a black hole at the centre of the Galaxy was somehow sending out powerful beams of gravitational waves. Believing the results to be true, astronomers began searching for visible signs that thousands of stars a year were being converted into gravitational waves. Experimentalists rushed to build copies of Weber's detectors to join this new field of astronomy.

In the early 1970s, Weber continued to report more signals. He found evidence that the waves were indeed from the centre of the Galaxy. But in 1972, other researchers began reporting their findings. Detectors set up by IBM Research Laboratories, Bell Telephone Laboratories, and laboratories in Paris, Munich, Rome, Glasgow and Moscow, all duplicated Weber's search. They found nothing. For them, the cosmos was silent.

Later that year, a one-day conference on gravitational waves was held at the Massachusetts Institute of Technology. Most of the new entrants to the field were there, and two key speakers were Tony Tyson of Bell Labs and Richard Garwin of IBM, in addition to Weber himself. Tyson and Garwin had been among the first to be disillusioned about the reality of gravitational wave signals and this was a huge disappointment to them. A gentlemanly professor on elbow crutches by the name of Phillip Morrison was chair. Weber presented his latest results and then Garwin spoke.

Garwin bluntly announced that Weber's results were invalid. Two embarrassing errors had come to light. One was a computer programming error that caused double-counting of some of Weber's data, thus creating false coincidences. The second was worse. Weber had exchanged data with one of the new groups. He had analysed the data and seen coincidences. Later it was revealed that one set of data had been recorded in

Greenwich Mean Time, while the other was using (American) Eastern Standard Time. With a four-hour discrepancy between the tapes, the coincidences were clearly spurious. On discovering the error, Weber had reanalysed the data and still saw coincidences, though fewer than before.

But Garwin was in for the kill. He didn't believe Weber's results and was determined to discredit him. Weber became very upset and ominously the pair became louder and angrier. They approached each other with clenched fists at the front of the lecture hall. As they appeared to be getting ready for blows, Professor Morrison hobbled over to them. When he raised a crutch between them, the tension began to subside, and the two men returned to their seats.

Other new detectors were also reported at that meeting, all similar in most respects to Weber's. However, in 1971 Stephen Hawking and Gary Gibbons had proposed a better way of reading out the signal from a bar. Weber had simply measured changes in the vibration energy of his bars. Hawking and Gibbons pointed out that a gravitational wave would not necessarily increase the ringing of the bar. Since the bar is always vibrating, a gravitational wave might equally act against the vibration and reduce its amplitude, or it might simply shift the phase of the bar's vibration.

Gibbons and Hawking realised that it would be easy to measure all of these different actions of a gravitational wave on a bar by measuring the bar in phase space. Their idea was to use the 'wagon-wheel' effect that makes wagon-wheels and other rotating objects look stationary in movies. The effect arises when a strobe light or successive camera frames are in synchronism with a rotation, so that a rotating object appears still. Using a carefully timed strobe light you can observe tiny changes in an object's rotation that you could never observe otherwise.

This same idea can be applied to electronic signals. By comparing the vibration of the bar with a strobe oscillator, it is possible to detect any changes in the vibrations of the bar—just like the wagon-wheel rotating faster or slower than the movie frames, causing it to 'rotate' forwards or backwards. By multiplying the voltage signal from the oscillating bar with the strobe oscillator, two signals are produced called quadratures. A graph of these quadratures (known imaginatively as X and Y!) is called phase space. A plot of X and Y in phase space produces an arrow called a phasor. The length of the phasor indicates the amplitude of the oscillating bar; the phasor's direction indicates the relation between the phase of the bar and the strobe oscillator. As the strobe oscillator is tuned to the frequency of the bar, the phasor rotates more slowly, just like the wagon-wheel in the movie. If the bar frequency differs from the strobe oscillator's frequency by one cycle per second, then the phasor rotates once per second: clockwise if the bar frequency is too high, anticlockwise if the bar frequency is too low. When the strobe oscillator has the same frequency as the bar, the phasor is steady like a wind vane, changing in length with variations in the amplitude of the vibration, turning slightly as the bar frequency or phase fluctuates.

Gibbons and Hawking suggested measuring the Brownian motion of the bar in phase space. When a gravitational wave strikes, the point of the phasor should suddenly jump. It may jump in any direction. To detect gravitational waves, you simply search for small jumps in the position of the arrow head (phasor).

This monitoring method was quickly adopted by the new detector groups, and by Weber. It should have been a much more efficient method for searching for gravitational waves. With a phase-space detection system in place, Weber should have seen even more gravitational wave events. He didn't. Others didn't see anything at all.

A final important development in our understanding occurred in the early 1970s. People began asking just how sensitive these resonant bar antennas could be. For years the gravitational wave research community failed to take into account a basic aspect of quantum mechanics: the uncertainty principle. This says that when you try to measure an object you exert forces on it, and these forces disturb what you are trying to measure. Usually the uncertainty principle is considered relevant only if the object is an electron or an atom.

To measure the motion of his bar, Weber had used piezoelectric crystals, amplifying the small signal using transistor amplifiers. We usually think of an amplifier as a device where you put a signal into the input and get a bigger signal out the other end. For example, the tiny squeaky sound of a record player stylus on a record is amplified into the ear-splitting, belly-shaking beat of loud rock music. But it also works the other way: shout into a loudspeaker and, believe it or not, a tiny but non-zero voltage will pass backwards through the amplifier and come out of the stylus as a tiny sound. This is part of a general rule: there is no such thing as a perfect one-way valve. Put another way, every action has an opposite reaction, not just in the mechanical sense of Newton's laws, but in every sense.

In the case of the Weber bar with the piezoelectric crystals, we must remember that the amplifier also contains resistors and these create electronic noise, in the form of voltage fluctuations. These voltage fluctuations can act on the crystals. The crystals themselves are near-perfect *reciprocal* devices: a displacement creates a voltage, and equally, (reciprocally!) a voltage creates a displacement. So the measurement system actually acts back on to the bar and makes it vibrate randomly. The amplifier puts noise back into the bar.

In 1975 Robin Giffard of Stanford University and Vladimir Braginsky of Moscow State University both showed that, no matter how ideal you make the measuring systems, whatever vibration sensor you use, you are ultimately limited in the measurement of gravitational waves by just this phenomenon. Even a perfect amplifier has an unavoidable backflow on the system being measured as a direct consequence of quantum mechanics.

It came as a shocking blow to realise that quantum mechanics actually limited the measurement of vibrations in tonnes of metal. It was always thought that quantum mechanics applied to atoms and not to macroscopic objects. Everyone overlooked the microscopic scale of the energy they were trying to measure. In 1975 the best detectors were 100 million times less sensitive than the limit set by quantum mechanics. They were limited by back action only because the amplifiers were far from ideal. But today the quantum limit is a very real barrier, but also one that might yet be beaten.

In 1987 an event occurred which should have generated detectable gravitational waves. The event was supernova 1987A in the Large Magellanic Cloud, the first supernova visible to the naked eye since 1604 (Chapter 6). At the time of the explosion, Weber's detector was still running, as was the detector in Rome. Various neutrino detectors, originally built to search for the radioactive decay of protons, were also operating. Three neutrino detectors—those in Japan, the USA and Russia—detected neutrinos which seem to have come from the supernova core collapse. A fourth neutrino detector in the Mont Blanc Tunnel in Europe also saw a burst of neutrinos, but these appeared several hours before the other neutrino signals. The trouble with interpreting these neutrino data arises because the neutrino signals come along with a constant background noise from radioactivity and cosmic rays. A single flash normally means nothing; you need several at once to be significant.

At a conference later that year, Weber and Guido Pizzella from the University of Rome reported the results from their detectors. Pizzella's detector had shown a modest 'event' close to the time of the Mont Blanc signal. The size of the signal was about the size he would expect to see randomly once a day, presumed to be the results of external vibrations making it through the vibration isolation system. Weber also reported a signal in his antenna, but his signal was a few seconds later than the Rome detection signal. Weber claimed this was confirmation of the event, and of his old data. This was difficult to accept, however, since gravitational waves travelling at the speed of light do not take several seconds to cross the Atlantic Ocean, but complete the journey in about 10 milliseconds.

Pizzella spent the next year examining all the neutrino data and gravitational wave data from the day of the supernova. He found an extraordinary result. There appeared to be coincidences between all the neutrino detectors and both of the gravitational wave detectors. The single neutrino events normally thought to be background radioactivity were in coincidence with small signals in the bars. This went on for a period of several hours around the time of the supernova; all up, there were some 100 coincident events. This was unbelievable. The total gravitational wave energy would have to be equal to an event involving 100 000 solar masses; the neutrino energy would also be enormous. The trouble with the analysis was that several corrections were applied to the data. These included corrections to the clocks at some of the detectors (the Kamiokande clock in Japan had a few seconds error) and delay times for the neutrinos, because they may have a non-zero rest mass and take a bit longer to get here than the gravitational waves because they travel at just below the speed of light.

The timing corrections reduce the significance of the coincidences of Pizzella and his many co-workers. However, even allowing for these, the coincidences are surprising. Unfortunately, there is no good theory with which we can interpret the data. The experiment cannot be reproduced with another nearby supernova. The signals cannot be gravitational waves as usually understood, and they cannot be neutrinos. But there is an enormous amount of dark matter in the Universe. Perhaps there are two presently unknown types of exotic particles, both of which are emitted in supernovae. Perhaps one of these particles interacts with resonant bars and the other (which travels slower or is emitted later) with neutrino detectors. But when a strange result requires two exotic explanations, you have to be extremely cautious. We simply have to wait until the development of more sensitive gravitational wave and neutrino detectors—and another supernova—before we can finally settle the question of the coincidences of neutrinos and gravitational waves.

There is a lesson to be learned in the apparently false alarm of Weber's discovery of gravitational waves: the importance of scepticism to the progress of science. Without it, science is dead. If you as a scientist believe too much in your results, you are in big trouble. All scientists, in fact everyone interested in the truth, should consider more carefully the power of scepticism. This is not a new idea. It was formulated 2600 years ago by Gautama Buddha in the *Kalama Sutra*:

> Do not believe anything simply because you have heard it . . . do not believe anything just because it is written in books . . . do not believe in what you have imagined . . . do not believe anything merely on the authority of your teachers or elders . . . but after thorough investigation

> and analysis, when you find that anything agrees with reason and is conducive to the good and benefit of one and all, then accept it and live up to it.

Although it seems that Weber's results fail the acid test of scepticism, he triggered the physics world's search for gravitational waves. Undoubtedly, the success of that search will come much earlier than it would have without his outstanding pioneering research.

Note

1 Vibrations, like sound and radio waves, can be measured in hertz, where one hertz is one vibration per second.

CHAPTER 11
SUPER DETECTORS

While Weber pioneered the search for gravitational waves, his detectors were not sensitive enough to detect them. Since his original 1960s experiments, better detectors have been developed which make use of superconductivity, a property that many materials demonstrate when cooled to very low temperatures. So far this approach has yielded a 10 000-fold improvement over Weber's original devices, and their successful operation represents a milestone in the search for gravitational waves.

A key player in the story of cryogenic resonant mass detectors—huge metal bars cooled to near absolute zero—was an American physicist, Bill Fairbank. Fairbank's childlike enthusiasm and joy in physics were infectious. He wasn't an eccentric; nor was he an intimidating intellectual. However, he did participate in the unfolding of many of the spectacular properties of superconductors and superfluids. During his last twenty years, he found ever-new ways of applying their properties to an enormous range of extraordinarily difficult, bold and exciting experiments. One was the relativity gyroscope experiment we looked at in Chapter 4: this was one of Fairbank's first projects at Stanford University.

The full story of Bill Fairbank is recounted in *Near Zero*,[1] a book from a conference celebrating his sixty-fifth birthday. It contains a detailed account of the relativity gyro experiment and of the experiments we will look at here. Bill Hamilton describes how he and Fairbank first became interested in gravitational wave experiments in 1967. Hamilton was finishing his thesis, taking data between three and five o'clock in the morning. Fairbank was always around and they often discussed Weber's experiments during these cold morning hours. According to Hamilton, Fairbank's rule was: 'An experiment can always be done better if it is done at low temperatures'. But very wisely, after long years of involvement in Fairbank's projects, Hamilton adds his own corollary: ' . . . but it is always much more difficult!'

We looked briefly at superconductors in Chapter 4 and saw how they can be applied to gyroscopes. Now let us take a closer look at superconductors and see how they help in the search for gravitational waves.

When many metals and alloys are cooled, they pass through a sudden transition into the state of superconductivity. The transition is similar to the transition from solid into liquid, or liquid into gas, and occurs typically at temperatures between a fraction of a degree and tens of degrees above absolute zero. The first material seen to enter into this state was the metal mercury. In 1911 the Dutch physicist Kammering Onnes observed the sudden vanishing of all electrical resistance in mercury at around 4 degrees above absolute zero (4 kelvin, or 4K). Many other pure metals—aluminium, tin, lead, niobium, tantalum, uranium, just to name a few—similarly pass suddenly into the superconducting state, each at its own characteristic temperature. In the superconducting state, electrical currents can flow forever. In a loop of wire or an electromagnet with the ends of the coil joined, the electrical current will flow without decay so that magnetic fields can be maintained without any external energy input. In magnetic resonance imaging machines, used for medical diagnosis, you are placed into a powerful magnetic field created by just such a persistent superconducting current.

Onnes was also the first person to liquefy helium. This provided the means of creating the low temperatures needed to study superconductivity. Although helium boils at a temperature of 4 degrees Kelvin at atmospheric pressure, it can be cooled even lower by reducing the pressure, in the same way that water boils at a lower temperature in the thin atmosphere on high mountains. It was surprising that Onnes did not notice immediately that, when helium is cooled below 2.17 degrees, its appearance suddenly changes. Helium's normal appearance, through clear gaps in the thermos flask containing it, is like clear, boiling water. Below 2.17 degrees, however, all the boiling suddenly stops, and it becomes quite still. It took fifteen years before this phenomenon was noticed, and a further twenty years before it began to be understood. This is surprising, because the change in helium was just as remarkable as the onset of superconductivity in a metal. Below 2.17 degrees Kelvin helium becomes a 'superfluid': its viscosity falls to zero, so that the fluid can flow without resistance. Its thermal conductivity also becomes enormous, so that heat flows through it with zero temperature difference. It was not until the 1950s that superconductivity was properly explained. Then it was realised that superconductivity and superfluidity are two manifestations of the same physical phenomenon.

Superconductors can demonstrate some bizarre and useful properties. For example, if you make a small, shallow saucer from lead and place a magnet on it, and then cool the saucer with helium, a spectacular phenomenon occurs. As the lead cools through about 7 degrees (its superconducting transition temperature), the magnet suddenly leaps to life and hovers suspended, quivering, above the saucer! This is a manifestation of a fundamental property of superconductors—they cannot tolerate magnetic fields inside them; magnetic fields are always expelled from any superconducting region. They do this by creating 'image currents' in a very thin surface layer. The image currents create a magnetic field which is the mirror image of the external magnetic field of the magnet. So the magnet 'sees' its mirror image in the superconducting surface and is repelled by it, in the same way that the matching north or south poles of magnets tend to repel each other. If the magnet moves closer to the dish, the mirror image grows stronger and pushes harder against the magnet. The magnet soon settles down into a state of mutual repulsion with its image, and hovers in perfect equilibrium above the dish. This property is a major factor in the way modern gravitational wave detectors work.

Before we look at the use of superconductors in the search for gravitational waves, let's pause to make a general comment about science, superconductivity, and on our particularly poor ability to predict natural phenomena.

Science prides itself on its predictive powers; when things are simple, it certainly succeeds. Examples of successful predictions are the deflection of starlight by the Sun, the existence of Neptune and neutrinos, and the possibility of nuclear power. But when things become more complex, predictions seem to miss their mark. Physicists may be forgiven for not predicting the bizarre phenomenon of superconductivity, but nonetheless it took fifty years to explain it once it was discovered, and thirty years later physics failed to predict so-called high-temperature superconductors, such as copper oxide superconductors, discovered in the 1980s.

In a similar way, every new branch of astronomy has yielded surprises. In the 1950s, when radio astronomers were planning their giant radio telescopes, their optical colleagues said, 'there will be no significant discoveries or sources discovered at radio wavelengths'.

Yet radio astronomers went on to discover pulsars, quasars and radio galaxies, discoveries that revolutionised astrophysics. X-ray astronomers fought a similar battle, but triumphed with their discoveries of black holes, accreting neutron stars and gamma-ray bursters. Today, many critics are claiming that the search for gravitational waves is a waste of limited resources. Some say there will be no sources. A few go so far as to suggest that gravitational waves may not exist.

They may be right about the sources. It is impossible to prove the potential discoveries that lie hidden. No matter how good the track record of unexpected discoveries, there is no way of proving that the same glory awaits gravitational wave astronomy. But using the superconductivity analogy, it is just as likely that an enormous variety of rich and subtle gravitational phenomena await discovery, much more intricate and revealing than any of the sources we have predicted. Still, one can be very sceptical—quite rightly so!—when observations seem to imply phenomena totally at odds with our existing knowledge. Signals that require the energy equivalent of 1000 solar masses for their creation (such as Weber's signals) were always barely credible. Yet Weber's research was probably essential to trigger the interest of people like Fairbank who said: let us use low-temperature techniques and build detectors a hundred million times more sensitive than Weber's.

Now let's return to our story of the gravitational wave searchers. Bill Fairbank and Bill Hamilton visited Weber in 1967 a few years before any of his dramatic announcements. In their late-night discussion they asked how cryogenic techniques could be used to create a detector much more sensitive than Weber's. Could superconductivity be used to make more sensitive vibration sensors? Could low-temperature techniques be used to reduce the thermal noise of the bar? Could magnetic levitation be used to isolate the bar against vibration?

The answers to all of these questions seemed to be 'yes'. The idea was to cool a 5-tonne bar to the lowest possible temperatures inside an enormous vacuum flask called a cryostat, suspend it magnetically using superconducting coils, devise superconducting vibration sensors, and so be able to detect realistic sources of gravitational waves from the expected formation of neutron stars and black holes. If they could achieve the sensitivity predicted, they would be able to examine Weber's gravitational wave signals in detail, tracking down their locations in the sky and correlating them with phenomena seen in the electromagnetic spectrum.

Many problems had to be solved to achieve Fairbank's goal. First, the temperature. If a bar was cooled from room temperature, 300 degrees Kelvin, to the boiling point of liquid helium at 4 degrees, its thermal vibrations would be reduced by a factor of 4/300. This reduces the noise significantly, but reducing the temperature to, say, 40 millidegrees above absolute zero (40mK) would reduce the noise by a further factor of 100. About a decade earlier, a device had been invented called a helium 3–helium 4 dilution refrigerator. This device made use of the cooling produced when the rare helium isotope helium 3 was dissolved in helium 4. It allowed samples to be cooled down to below 10mK. But these complex gadgets had been used to cool only small samples. Could they be modified to cool a lump of metal weighing 5 tonnes?

Fairbank made this bold suggestion, knowing that the heat capacity of metals falls drastically as the temperature falls to absolute zero, and that, as long as the thermal insulation was sufficient, the enormous mass did not represent a serious problem. As it turned out, Fairbank was right! It was done by Guido Pizzella and his colleagues at the University of Rome. However it was twenty years after Fairbank's original suggestion before this was achieved. Sadly, by this time, Fairbank had died.

For their first decade of operation, cryogenic gravitational wave detectors had to settle for 4 degrees operating temperature. Even achieving good operation at this much easier temperature has been an extraordinarily difficult task.

Fairbank's agenda consisted of three goals: ultra-low temperature cooling, superconducting vibration sensing, and magnetic levitation for vibration isolation. Only the last of these was never achieved, and has now been abandoned. The problem with levitation is that better and better superconducting properties are required as you increase the mass being levitated. For small bars, it works very well. You wind 'pancake coils' consisting of single-layer spiral coils, which are mounted on the surface of a metal cradle. The bar must be superconducting. Aluminium, which is cheap and the traditional material for resonant bars, becomes superconducting only below about 1 degree, and even then it has a low critical magnetic field. If a magnetic field above the critical magnetic field is applied to a superconductor, it reverts to the normal state.

At Stanford University, extensive experiments were performed using a method called 'plasma spraying' to spray a high-quality superconducting alloy onto the surface of aluminium bars. This worked well enough to allow small bars to be levitated, but for big bars the maximum possible levitation force was insufficient. At a magnetic field well below the critical magnetic field, the magnetic image collapsed, and the field penetrated the superconductor, destroying the magnetic repulsion.

About 1970, Bill Hamilton moved from Stanford to Louisiana State University. At Louisiana, Hamilton's group tried gluing thin superconducting sheets around a 1-metre-diameter aluminium bar. In this case the bar could be levitated, but the glue joint ruined the acoustic properties of the antenna.

The University of Western Australia gravitational research team have a bar made from niobium which is the best superconducting pure element. At the Perth lab, it was hoped that the better properties of niobium would allow simple levitation. While the critical field was known, it was not known whether the magnetic field would penetrate at a lower field. Nor was it known how strong magnetic fields would degrade the acoustic properties of the bar. The first experiment was a tiny 5-centimetre-diameter antenna, 30 centimetres long. This was followed by a larger version, a 1-metre-long bar, 10 centimetres across. In both cases the levitation worked beautifully. A set of coils lifted the bars to a height of one centimetre, where they floated freely, completely isolated from external vibrations. Their acoustic losses were also very low—typically, they rang for many hours after being struck—just what was needed for a sensitive detector. Tests were also performed on small niobium disks, directly measuring the levitation force achievable. Here was bad news: measurements revealed an upper limit on the size of the niobium bar that could be levitated, and this was only at 20 centimetres diameter.

At this stage, the Perth team had to make a decision: go ahead with levitation and sacrifice sensitivity by keeping the bar's diameter small, or abandon levitation and buy the biggest bar they could afford? They chose the latter. In the short term, this turned out to be a mistake. The experimental chamber of the cryostat was not well enough insulated against vibration to cope with a non-levitated test mass. The team was unable to afford the cost of a major reconstruction; patch-up after patch-up was attempted, all unsuccessful. It was ten years before the antenna operated as well as the original baby bar, but by then the niobium bar experiment was the best in the world. Similar problems had dogged other research groups, all ultimately a result of the failure of magnetic levitation.

Fairbank's goal was to make detectors capable of sensing gravitational wave bursts with an amplitude of one part in 10^{20}. This means that the gravitational wave would change the spacing between objects, or the length of a bar, by one part in 10^{20}. For a bar of a few metres in length, this represented detecting vibrations of 10^{-20} metres! Could such small motions ever be detected, even in principle? These vibrations were 10 000 times smaller than anything Weber had measured, and corresponded to 100 million times less energy. It was a motion as small compared with an atom, as an atom is compared with a human being. Does it make sense to measure such small distances? At one level, this question of scale is easily answered. The size of the atoms does not really matter because the measurement is the average distance, averaged over a surface containing perhaps 100 trillion atoms. In a similar way, Earth-orbiting satellites can measure the sea level in the oceans to an accuracy of about one centimetre, even though ocean waves are hundreds of times larger.

At another level, the question is much more profound. Surprisingly, as we noted in the previous chapter, nobody had asked about the influence of quantum mechanics. No one had imagined the measurement of vibrations in tonnes of metal was anything but a classical process. Quantum mechanics applies to atoms and electrons, not things as big as a car! But the realisation that bars were quantum objects greatly enriched the field, led to some wonderful physics, and challenged us to be much cleverer than we ever imagined was necessary.

Any motion sensor that is supposed to be very, very sensitive must have very low resistance because resistance always creates electrical noise. Immediately you think of superconductors, materials with zero resistance. Fairbank gave a Korean postgraduate student, Ho Jung Paik, the task of devising a superconducting sensor. Such devices are usually called transducers. Paik came up with a very elegant solution. First he asked: how can you increase the size of the vibrations? Answer: join the large resonant bar to a much smaller resonant mass. To understand why this is important, try a simple experiment. Make a heavy pendulum, anything will do—a brick on a string, a can of beer and a shoe lace, or swing an electrical appliance from its power cord (not recommended!). Now from this pendulum, hang a second, the same length but much lighter. Try a cotton thread and a grape, or a stone or coin. Now you have a double pendulum. Give the top, heavier pendulum a push and watch what happens. For a moment, the smaller pendulum remains still, but it soon starts to move and eventually builds up a much larger swing than the heavy one. After a while the heavy pendulum will stop swinging altogether and the smaller one will be swinging wildly. After a while, if your pendulum is any good, the small one will stop and the larger one will start again. What you have just observed is the energy transfer from the heavy resonator to the lighter one, then back again, and so on. Paik arranged for the bar's motion to be transferred to a thin superconducting diaphragm in the same way.

The next problem was how to measure the motion of the diaphragm. Paik placed coils on either side of the diaphragm and linked them to a SQUID (a superconducting quantum interference device, which we first met in Chapter 4). SQUIDs use interference of superconductivity 'wavefunctions' to create supersensitive current meters. It wasn't long after SQUIDs were invented that they were available commercially, and so Paik's vibration sensor sounds simple: a diaphragm, two coils and a SQUID. The reality is far from this. An enormous number of technical problems had to be solved to make them work well. Wires broke when the coils were cooled to helium temperatures and there were electrical losses in the wires. All up, it meant years of tedious experimentation.

By about 1980 the Stanford group had demonstrated good results using Paik's sensor on a bar at 4 degrees. They had achieved about a 1000-fold increase on the sensitivity of Weber's detector but, unfortunately, no sign of gravitational waves. Eventually the groups at Rome and Louisiana adopted variants of this technology. Better SQUIDs were invented to gain further improvements.

In the meantime, groups of physicists from Moscow, Rochester, Rome and Perth persisted with a totally different superconducting vibration sensor called a 'parametric transducer'. Parametric transducers are vibration sensors that convert the vibration of the bar from its actual frequency to an arbitrary high frequency, one chosen because it is easy to amplify at that frequency without adding too much noise. A radio transmitter converts sound frequencies to the radio frequency so that it can be broadcast as a radio wave. Likewise the parametric transducer converts the vibration of the bar to a microwave frequency (the frequency used in radar). But in a radio studio the sound frequency is turned directly into an electrical signal (that is what microphones do) before impressing the signal onto a radio wave. Transducers convert the vibration of the bar to microwaves without any intermediate stage. The conversion is done by a nearly lossless superconducting resonator so as to add the minimum possible noise to the measurement. Today sensors of this type have proved a bit better than SQUID devices, but a friendly rivalry is still under way between half a dozen groups all aiming to outdo each other.

Once detectors were capable of registering incredibly small vibrations, there remained the problem of how to isolate the resonant bar antennas from external sources of vibrations. When magnetic levitation failed, another method had to be found. Cryostats are awfully noisy environments, full of creaking pipes and metal plates, and boiling liquid air and liquid helium. They must

Figure 11.1 A resonant bar gravitational wave detector as realised at the University of Western Australia. The bar is cooled to a few degrees above absolute zero and isolated from vibration. The transducer uses microwaves and superconductivity to read out the vibrations of the bar.

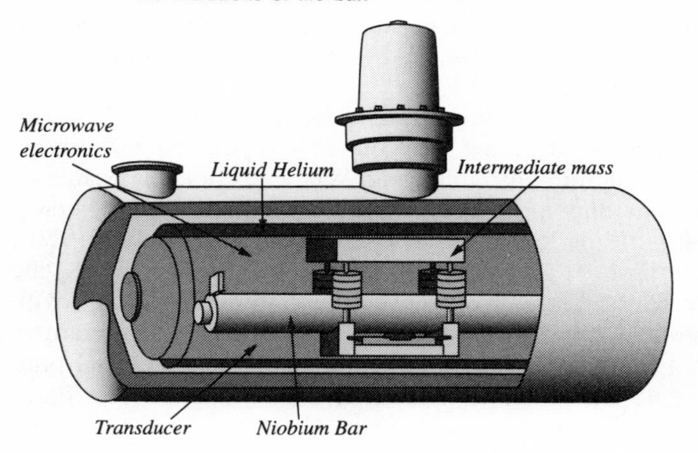

support very heavy loads, using the thinnest possible cables to prevent unacceptable amounts of heat entering from the outside world. This makes design very difficult. Let's look at how the Perth design solved these problems.

A car's suspension is a single-stage *vibration isolator* consisting of one mass (the car), one spring (per wheel) and a damping mechanism (the shock absorber). You can increase the isolation by putting many such systems in series. For example, mount your car on the back of a small truck, mount it on a bigger truck, and put the whole lot on a huge truck, and you will have a ride which is softer and quieter than money could buy in any single car. In the Perth design about 17 stages are used: some consist of lead disks separated by rubber blocks, operating at room temperature to keep the rubber soft. These are followed by steel masses and cantilever springs, not unlike car springs, in the low-temperature environment. Finally, the antenna itself sits on a carefully designed, tapered titanium spiral spring called a 'catherine wheel'. With this system you can hammer the chamber containing the bar and drop bricks on the floor without disturbing the peace of the antenna. No wires are attached to the bar: all signals are sent in and out in radar-type microwave beams, so no vibration enters down wires.

The heart of the resonant bar antenna is, of course, the bar itself. How do you choose the bar? Fairbank chose aluminium, following Weber's choice, because it is light and inexpensive, and has a high quality factor. This means low losses and less noise. In the late 1970s, the Tokyo gravitational wave group led by the late Professor Hirakawa discovered an aluminium alloy, with about 3 per cent magnesium, which at low temperatures had a quality factor ten times higher than the usual pure metal. When one of us, David Blair, (leader of the Perth team) was at Louisiana State University, he discovered that niobium has a very high quality factor. The Perth group, as we have seen, chose niobium after several years of intensive study. The niobium bar they eventually purchased cost a quarter of a million dollars, but turned out to have a quality factor many times better than the aluminium bar used by the Tokyo group. With such low losses—the bar actually rings for several days once it starts—the noise level at 4 degrees is very low and the sensitivity is limited by the sensitivity of the transducer.

The bar shape was chosen purely for convenience. In general, a bar needs to be as heavy as possible, but still with its resonant frequency close to the desired frequency—say 500 to 1500 hertz, typical of the most likely gravity wave sources, such as supernovae. Some groups are planning to build enormous spherical antennas weighing between 30 and 100 tonnes. This gives extra sensitivity due to the mass and an added bonus of omni-directional sensitivity. A sphere is perfectly matched to the deformation patterns produced by gravitational waves, as long as it has transducers in various locations to detect the vibration pattern. The latter is a distinct advantage over bars, which are insensitive to waves coming in along their axes.

Two groups in Italy, at Rome and Padua, are working on *ultracryogenic* antennas. In Rome they have already cooled a big bar to about 80mK, knocking down the thermal noise of the bar by a factor of 50 compared with bars at 4K.

Fairbank's dream has slowly become reality. After twenty-five years of painstaking efforts, three antennas around the world are in almost continuous operation, waiting and hoping for a supernova or an unseen gravitational collapse at the heart of our galaxy. The goal is to see three nearly simultaneous bursts from these detectors. There is a negligible chance that three signals will happen simultaneously by accident, so a coincidence will set alarm bells ringing. The chances are small because supernovae are not common in our galaxy, but the instruments are so clean and reliable that they offer the best chance this century for detecting the first gravitational wave signal.

Let us now come back to the question of quantum mechanics, and in particular the 'quantum limit'. This is the fundamental limit to the level of detectable vibration that no one, at first, wanted to acknowledge. It was first publicised by a Russian academician, Vladimir Braginsky. Braginsky has a small, cluttered office in the low levels of Stalin's enormous wedding-cake castle known as Moscow State University. Braginsky looks Russian: round face, jovial smile, and always in need of a cigarette.

Your first encounter with Braginsky is always tense. He talks in riddles; he avoids conversation. The talk moves in circles as if there is some sort of mistrust. This ritual is formalised through the brewing of tea: a Bunsen burner, a round chemical flask and enormous quantities of tea leaves create a bitter brew that cannot possibly be drunk without massive amounts of sugar. After tea, there are more riddles. You feel stretched, you are reprimanded for not reading his papers properly. Mysterious, new ideas emerge as you scribble notes, hoping to make sense of it all later.

Braginsky has always believed that gravitational waves should be detected more by cleverness than by brute force. Brute force means enormous masses, huge-scale detectors and ultralow temperatures. The brute force approach is difficult, inelegant and very expensive. Braginsky asked why can't we do it elegantly? Why should nature be so unkind and force us to go to such lengths to detect her subtle but energetic gravitational waves?

Braginsky was worried about the thermal noise of the resonant bars. His solution was to find a much better material: pure, artificial sapphire. This material has exceptionally low losses, low enough to kill all thermal noise. Braginsky proved this in the primitive conditions of Moscow State University. He persuaded a crystal-growing institute to make him a half-metre-long bar. In his words, you must 'hang it from the finest Chinese silk, or the finest Swedish tungsten'.

There were two problems with Braginsky's sapphire bar. At a half-metre in length it had a frequency of 40 000 hertz, ultrasonic and too high for likely gravitational wave sources. To get down to more probable frequencies a much larger bar is needed, but the cost is horrendous: one American company offered to attempt the manufacture of a 5-metre-long sapphire bar, 1 metre in diameter. There would be no guarantee of success, and it would cost $50 million.

The second problem relates to the quantum limit. Small sapphire bars have a more serious quantum limit to their sensitivity simply because of their low mass. We saw in the previous chapter how any amplifier can act back on what it is amplifying. A vibration sensor is a special type of amplifier, one which has a mechanical input and an electrical output. Mathematically, mechanical and electrical vibrations have a remarkable similarity, like the similarity between water waves and electromagnetic waves. The similarity means that we can treat the vibration transducer which senses the motion of the bar, and the amplifier which follows it, all as a single unit, a single amplifier. Like all amplifiers, however, it amplifies the motion but also acts backwards, exerting noise forces onto the bar.

These noise forces cause fluctuations in the bar exactly like thermal noise forces. Even in a sapphire bar, with its high quality factor and no thermal noise, there remains noise which is like thermal noise but which is due to the measuring system acting back onto the bar. Even if the amplifier is perfect and the bar is intrinsically noiseless, this creates an unavoidable noise level equal to one quantum energy unit: Planck's constant times the bar frequency. This result was discovered for amplifiers in the late 1950s. It is not surprising that it should apply to bars. What is surprising is that it took so long to be recognised by gravitational wave researchers.

This quantum unit of unavoidable noise energy means that there is a limit on the smallest possible displacement measurable in the bar, about 10^{-21} metres, and it corresponds to a minimum gravitational wave amplitude of about the same value, depending on the length of the bar. This is called the standard quantum limit. The smaller the mass of the bar, the worse the quantum limit, so small bars of sapphire seemed to have problems.

Both Braginsky and Robin Giffard from Stanford University recognised this problem in 1975. Although no one wanted to believe it at first, recognising and understanding the problem quickly led to a solution (in theory at least). It also led to an awful Russian name to describe the solution: quantum non-demolition, or QND. Many physicists erupted in consternation on hearing about QND. 'You can't do that! It is violating quantum mechanics!' And yet they were wrong. It can be done, although so far no one has managed to apply it to gravitational wave detectors.

The improvement in resonant bar gravitational wave detectors has been, on average, about one order of magnitude per decade. Extrapolating, it may take another twenty or thirty years or so before the quantum limit is broken. When this is incorporated in the big new spherical detectors planned in Holland and Italy, we will have detectors able to detect gravity waves from super novae and black hole formation in the nearest 1000 galaxies. Meanwhile, resonant bar antennas are constantly on the air, waiting for rare but powerful gravitational wave events. The QND techniques invented by Braginsky offer opportunities for big improvements in the future. While these resonant bars listen to the cosmos, however, there is a completely different technology which offers an exciting alternative means of detection; it particularly promises to detect coalescing neutron stars in very distant galaxies.

CHAPTER 12
SHEDDING LIGHT ON GRAVITATIONAL WAVES

Figure 12.1 A laser interferometer gravity wave detector.

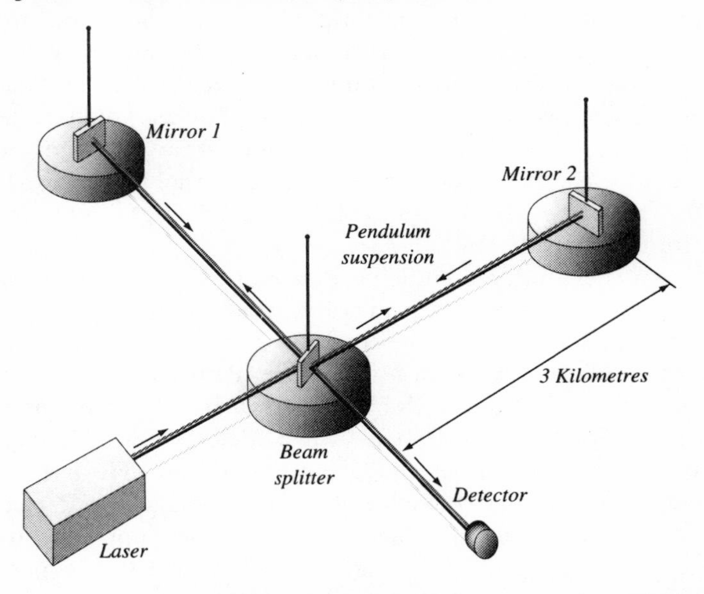

The idea of a laser interferometer gravitational wave detector. The beam splitter divides the light between two paths, one which shrinks while the other stretches as a gravitational wave passes.

During the 1960s, while Weber was experimenting with various resonant mass detectors, an alternative scheme for gravitational wave detection emerged—interferometry. The idea was to use a Michelson interferometer, not unlike the one used by Michelson and Morley to prove the non-existence of the aether, as we saw in Chapter 1. If you can imagine an interferometer with its two arms at right angles, the quadrupole deformation of a gravitational wave would cause one arm to shrink and the other to stretch. Remember that the light takes two alternative paths, and finally it recombines at the output. It recombines either with the two light waves adding or cancelling. It takes just one wavelength of light change in one path relative to the other for the light intensity at the output to vary from bright to dark. Thus a gravity wave would be detected as a change in brightness of the interference fringe on the output. To detect gravitational waves in this way, the entire interferometer would have to be suspended and isolated from vibration. The air would have to be removed and a very bright laser used, to make the greatest possible change in light intensity from bright to dark.

Weber and his students had considered the idea of using an interferometer in 1964, and unknown to him two Russians, M. Gertsenshtein and V. I. Pustovoit, had proposed the same idea two years earlier. Some years later, Weber's student Robert Forward built the first laser interferometer gravitational wave detector at Hughes Research Laboratories in Malibu, California. He used a helium neon laser and was able to show that the device could work, but that high sensitivity was not easy to achieve. About the same time that Forward was experimenting, Rainer Weiss at Massachusetts Institute of Technology set about a thorough analysis of the laser interferometer gravitational wave detector. Unfortunately he never published his analysis but merely submitted it to the National Science Foundation with a request to fund an experiment. They refused.

Meanwhile, the disappointed physicists who had set up resonant bars to check Weber's results were wondering what to do next. Some abandoned the field. For those who stayed, the choices were either to follow the Fairbank agenda and develop cryogenic bars, or to change to the laser interferometer approach. The group in Rome, led by Guido Pizzella, opted for the cryogenic bars, while the groups in Garching (near Munich) and Glasgow were inclined towards the interferometric approach. Just at this time the National Science Foundation sent out Weiss's proposal to be reviewed. It arrived on the desk of Herr Professor Billings in Garching. Billings was very excited and very enthusiastic about Weiss's proposal. He rang up Weiss and said he would like to pursue Weiss's plan in Germany. Weiss was livid, especially since the National Science Foundation refused to fund his project. Sometime later, Ron Drever's group in Glasgow started to follow the same path.

In the early excitement when gravitational waves seemed to have been detected, Drever and his colleague Jim Hough had built a bar detector consisting of two separate half-bars which sandwiched piezoelectric crystals between them. This gave good coupling of vibrational energy to the crystals, but the bad acoustic properties of the crystals also created a severe noise problem. Drever's first step into optical detection was to replace the crystals with a laser interferometer.

Because the wavelength of light is small—typically half a micrometre—an interferometer is intrinsically sensitive. At the very least it can measure one wavelength of light. It can easily measure this half a micrometre of movement regardless of how far apart the mirrors are placed. Yet in terms of gravitational waves half a micrometre is not small at all: we need to measure a trillion times less than this.

Of course, if you can measure one wavelength, you

can also measure one-tenth of a wavelength, or one-hundredth, perhaps one-thousandth. Just how far you can go in subdividing the wavelength of light was analysed by Weiss. You can think of the wavelength of light as the basic measurement of length, just as a 30-centimetre ruler is the basic measurement of length in the classroom. How finely you can subdivide your rule depends on a combination of ingenuity and fundamental limits. On a ruler, the width of the lines cannot be less than the size of an atom. It can't even be less than the wavelength of light used to read the lines, otherwise it would not be possible to discern them. Even close to this size, they would be invisible without the use of a microscope. But an ingenious experimenter will choose the shortest practical wavelength and the best possible microscope. Weiss applied this sort of analysis to a laser interferometer and showed that with a few clever tricks and very powerful lasers it was possible, in principle at least, to reach sensitivities better than the resonant bars.

Guided by Weiss's analysis, the two dedicated teams of brilliant experimental physicists at Glasgow and Garching set to work to explore the limits of laser interferometers. The question was, what do you have to do in practice to gain the necessary factors of a hundred million million from a single wavelength to achieve the desired sensitivity of one part in 10^{20} detectable change in length.

As a first step, Weiss had suggested using an 'optical delay line'. This involved replacing the single light path in an interferometer with a multiple path using a pair of curved mirrors. The laser beam can be arranged so that it enters the delay line through a small hole in one of the mirrors. The laser then bounces back and forth many times between the two mirrors before finally coming back out of the same hole. Suppose you have 100 bounces in each arm of the interferometer and then you recombine the beams. Such an arrangement is 100 times more sensitive than an instrument with only one bounce. The motion needed to create a change of one wavelength would be 100 times less, and the task of measuring the tiny vibrations 100 times less difficult.

The idea of using multiple bounces is not new. Weber had realised that the sound in a resonant bar antenna had to be able to sustain multiple bounces—hundreds, thousands, even millions of bounces—to reduce the thermal noise of the bar. Bill Hamilton had realised that the electric current in a parametric vibration transducer had to bounce back and forth a similar number of times to increase the sensitivity of the transducer. In these systems the number of bounces is called the quality factor, which is a measure of the smallness of the losses. But you can also think of the idea of multiple bounces as multiplying the sensitivity. The optical delay line, with its multiple bounces, has one difference from the high quality factor approach used for the bars—the multiple light paths are separate. In practice, however, the separation of the paths is quite unimportant. You can easily make the light reflect back and forth on itself, exactly like the resonance in a bar.

This was realised by Ron Drever. He chose to use a pair of very high reflectivity mirrors, one of which has partial transmission. You shine a laser beam onto the partially transmitting mirror—it might have 99-per-cent reflection and 1-per-cent transmission. The second mirror might have near-perfect reflection. When the spacing between the mirrors is exactly right, so that the reflected wave is exactly in phase with the incoming wave, the light builds up between the two mirrors until it is about 100 times as bright as before (like 100 superimposed beams in the delay line). Such a mirror configuration is called an optical cavity.

This is a stunning phenomenon. You shine a laser beam through one mirror to another. With a tiny adjustment of the mirror spacing, superintense light suddenly appears between the two mirrors, even though the incoming light is quite dim. In the very bright light resonating between the two mirrors you see the enormous concentrations of dust even in the cleanest air, and see the forces of radiation driving the sparkling dust particles along the beam.

If the spacing between the mirrors is not exactly correct, the appropriate wave additions and subtractions do not take place, and the light is 99-per-cent reflected off the input mirror (just as you would expect) and hardly any light builds up between the mirrors. If you graph the brightness of the light inside the cavity against the spacing between the mirrors, you will see the brightness low most of the time, but showing bright peaks whenever the spacing between the mirrors reaches the point where the light waves add up together to create a resonance. For exactly the right spacing between the mirrors, the intensity is enhanced 100 times, and this means that its sensitivity to changes in the spacing is increased by the same amount. This beautiful concept means that you can use a simple pair of mirrors to create a much more sensitive interferometer.

As with resonant bars, quantum mechanics places a fundamental limit on the sensitivity of laser interferometers. To start with, the uncertainty principle places a limit on measurements of the mirror positions by any means at all. You want to measure the relative positions of mirrors to an accuracy of 10^{-20} metres, but the measurement by an intense beam of laser light applies a force called radiation pressure. The photons in the light are like little bullets, and their impacts cause the mirrors to move. That is, the light impacting creates a momentum uncertainty. The limit that this sets is just like the quantum limit for resonant bars. However, if the mirrors are moved further apart, the size of the position uncertainty relative to the baseline is reduced. Bars cannot be more than a few metres long and the quantum limit corresponds to a fractional change in spacing in the range of 10^{-21} to 10^{-22}. The measurement using light allows the baseline to be, say, 1000 times longer, so the quantum limit to the gravitational wave strain is 1000 times smaller (depending, of course, on the mass of the mirrors and the signal frequency).

The second fundamental limit is associated with the quantum nature of light. Your laser may have a power of 1 watt, corresponding to about 10^{18} photons per second, or 10^{15} photons per millisecond. If you want to measure gravitational waves at 1000 hertz (1000 times a second), you need to measure changes occurring in less than one millisecond, that is, every thousandth of a second. But the intensity of any light beam always fluctuates unpredictably, due to the statistical nature of quantum physics: you can only specify the average number of photons. This statistical limit is set because there are only a finite number of photons, even though that number could be 10^{18} per second. You can use very simple statistics to calculate the limit set by the finite number of photons available.

The fluctuations in the number of photons obey statistical laws similar to those experienced by public opinion researchers. If you poll 1000 people, the statistical uncertainty is equal to the square root of the number of samples. That is, the square root of 1000, or just over 30. That means the uncertainty of the opinion poll is roughly 30/1000, or about 3 per cent. Most political opinion polls have about this accuracy, which often is not enough to predict election results. To improve the accuracy to 1-per-cent uncertainty would require the number of samples to be increased to 10 000. But this might cost ten times as much, and newspapers are unlikely to pay so much more for that extra accuracy. Besides, extra accuracy limits the scope of journalists to speculate and reduces the excitement of the election, which would be bad for newspaper sales!

In the case of our 1-watt laser, we have 10^{15} photons per millisecond, which has a square root of 30 million. This means that the intensity of the light will be observed to fluctuate by three parts in 100 million. This fluctuation is unavoidable, just like the 3-per-cent uncertainty in the opinion poll; it means, roughly, that you can never subdivide a wavelength of light to a fraction smaller than three parts in 100 million for the particular laser power that we considered. Most measurements with laser interferometers usually boil down to counting photons.

When you put in the numbers, you get some frightening results. For a laser interferometer to achieve a sensitivity of 10^{-21} you need not 10^{15} photons per measuring time as produced by our 1-watt laser, but more like 10^{20} photons. That means a 100 000-watt laser! Not only that, but you need a baseline several kilometres long. The detection of gravitational waves is no longer a laboratory experiment. Now it calls for a major observatory.

Visions of this scale are difficult to sell. Mixed up with the vision of gravitational wave astronomy are power, egos and money. One Australian optical observatory director accused gravitational wave researchers of being 'a mob of physicists trying to hijack astronomy's money!' The trouble is that gravitational wave observatories are expensive, and some people imagine pots of gold reserved for their pet project in government treasuries.

After years of spending government funds on their own projects, some people on advisory committees suddenly develop scruples about the money that will be required for a project that is not their own. Of course, the fact that the leader of the rival project long ago won the heart of a woman they fancied, or was suddenly promoted past them, never affects their decision. In opposing the project, such people might 'accidentally' leave out a crucial piece of information that makes the proposal seem unlikely to succeed or even unnecessary!

Like any group of people, the physics and astronomy community is capable of such intrigue and infighting.

In the United States the battle to win funding for gravitational wave observatories was not helped by a mixed marriage, a difficult birth, an unhealthy childhood and a stormy adolescence. The story will provide fodder for future historians of science. Here are the briefest details.

The National Science Foundation eventually funded Weiss to begin development of laser interferometers at MIT. Kip Thorne, Professor of Theoretical Astrophysics at Caltech, persuaded his university to establish a laser interferometer gravitational wave program. They hired Ron Drever from Glasgow. At first he commuted between Glasgow and Los Angeles, but eventually he cut his ties at Glasgow. Jim Hough became leader of the Glasgow team and built it up to be an outstanding and dedicated group.

Ron Drever is a little, tubby, fast-speaking Scotsman with a wealth of ideas and an intuitive understanding of physics. He also has the ability to improvise and find uses for ancient pieces of junk. He used to incorporate radio-controlled toy cars in his experiments, and probably Meccano and train sets too. One of his more eccentric ideas was rubber toy cars sandwiched between steel or lead masses for use in vibration isolators, an idea that worked well.

Many of Drever's ideas have been of major significance. One of them was his invention of power recycling. We have seen that the sensitivity of an instrument depends on the maximum light intensity inside it. However, the signal from the interferometer is seen as changes in the brightness of the dark fringe. That is, when operating correctly, very little light enters the photodetector. Apart from other considerations, if all the laser power were to strike a photodetector, the detector would be destroyed. But energy is conserved; bright light has to escape somewhere. It escapes through the input port of the beam splitter back into the laser.

A true Scotsman, Drever said, 'Why waste this light? Why don't we recycle it?' He proposed sending the outgoing light straight back by placing a highly reflecting mirror at the input. Surprisingly, it works! When the recycling mirror is carefully positioned, the wasted light adds to the light already in the interferometer so that the total stored light can be built up many hundreds of times. This idea is now a cornerstone of laser interferometer technology.

Drever arrived at Caltech full of enthusiasm. He expected to be able to build a big detector, to be called the Laser Interferometer Gravitational Observatory (LIGO), in the next few years. It would be huge and it would cost a lot of money. Simultaneously, Weiss wanted his big MIT interferometer. Unfortunately, all went wrong. First, the National Science Foundation wanted MIT and Caltech to propose a joint project. This didn't work out. A caricaturist would portray Weiss as a tight, tough and rigorous professional, while Drever would be a loose, wild eccentric Doctor Doolittle. Both were brilliant, but so different they could never agree on anything. Eventually it was decided that an overall project director was needed. They hired Professor Robbie Vogt, a tough post-war German immigrant. Vogt was a member of the first tractor team to the South Pole, once Provost of Caltech, and once Chief Scientist at the Jet Propulsion Laboratory. His role was to create discipline and order and make the project happen. He secured major funding from the National Science Foundation. But still there were problems. Vogt eventually fired Drever from the project in 1992. That should have been the end of the stormy adolescence. But opinions were divided, and appeals brought a new round of infighting. Late in 1993 the President of Caltech dismissed Vogt. Things are best viewed from a distance, and so it is better left to history to decide the correctness or incorrectness of these decisions.

Robbie Vogt had created a tough, disciplined team of scientists and engineers. Everything was directed towards the goal of building two 4-kilometre by 4-kilometre laser interferometers. In the first instance, the instrument was designed to achieve a bit less than 10^{-20} strain sensitivity. Future improvements should increase the sensitivity a hundred times. At this sensitivity, such a detector should be able to detect gravitational waves from the coalescence of neutron stars.

To achieve such goals requires enormous advances in the technology of ultra-low-loss systems. Teams in France, Italy, Glasgow, Germany, Australia and Japan are all working on the problems. These problems are so difficult that co-operation and collaboration are far more important than competition, although there is still plenty of friendly rivalry and rigorous criticism.

The first problem that must be solved in making a sensitive interferometer is to eliminate the adverse effects of the air. It turns out that even an atmospheric pressure one-billionth of that we experience at sea level is far too high. To get down to such low pressures, very high performance vacuum pumps are needed. But for a large-scale interferometer you need 6 to 8 kilometres of pipe. Furthermore, the pipes have to have a fairly large diameter. The problem is that a laser beam does not remain a tiny pencil beam: it expands into a cone at a rate dependent on the diameter of the optics used. (Over 4 kilometres you need maybe 50-centimetre-diameter mirrors.) As the laser passes through the pipes it spreads a little, so that in smaller pipes it might reflect off the sides and back into the beams. Any vibration in the pipe would be carried by the reflected laser and interpreted by the system as a gravitational wave. So the minimum diameter of the pipes is about a metre. Over the kilometres of pipeline, this amounts to a huge volume to be evacuated.

To achieve a high vacuum over such a long pipeline you need to construct the pipe out of stainless steel. Then, given its enormous size, most high vacuum specialists would expect you to use a thousand vacuum pumps. The price of the pumps alone would be outrageous, so the laser interferometer projects proposed to achieve the same vacuum with a single pumping system. This meant eliminating the worst source of vacuum degradation: hydrogen gas dissolved in the stainless steel and slowly diffusing through the pipe.

The LIGO scientists decided to study the problem. Vogt persuaded a retired engineer, Baude Moore, who had spent his life working with vacuums, to come and work with them. Baude made scaled-down vacuum systems, which produced terrible hydrogen outgassing. He eventually found an easy solution, however: make the pipes out of stainless steel that has been baked in air at about 440 degrees Celsius—hot enough to tarnish it—for about 36 hours. This appears to oxidise the hydrogen, in other words turns it into water. Thereafter the hydrogen problem is overcome and now you can build a very cheap vacuum system.

The next problem with the laser detectors is the mirrors. Needing a radius of about 4 kilometres to help keep the laser beams tightly focused, these mirrors are incredibly flat, and yet the shape of the mirror has to be very accurate so that the wavefronts of the beams combine to form a dark fringe for interference. This means that the surfaces of the mirrors have to be accurate to about 0.1nm or one-tenth of a nanometre (a nanometre is a billionth of a metre). These are atomic dimensions. The mirrors must also have extremely low loss to allow the build-up of extremely bright light in the optical cavities.

When the LIGO-type detectors were first proposed, mirrors with losses of about one part in 10 000 were available. This was enough of a loss, however, that people imagined tens or hundreds of lasers would have to be combined to obtain enough laser power. But during the 1980s and 1990s, better mirrors have become available. Now you can buy super-mirrors that reflect 99.9999 per cent of the incoming light. The Australian team has shown that making the mirrors out of sapphire instead of normal silica glass can dramatically improve the sensitivity of laser interferometers. This arises for the same reasons that sapphire would be a good material for resonant bar detectors. This provides a hope that useful detectors can be made that are only half a kilometre in size, and much less costly.

This meant that it was necessary to construct sapphire mirrors. A team headed by Chris Walsh at the CSIRO's Division of Applied Physics in Sydney set about developing new surfacing techniques to polish sapphire mirrors. Their success has earned them a world-wide reputation. What sets these mirrors apart from precision telescope mirrors is they have to be extremely smooth across the entire surface. While ripples can be tolerated on an ordinary telescope mirror, the way in which the mirrors have to perform in gravity wave detectors means they have to be roughly 1000 times more perfect in both shape and reflectivity.

Artificial sapphire is made in 30cm size blocks of single crystal, simply by melting and carefully cooling alumina at about 2000 degrees Celsius. Sapphire being so much harder than glass meant that Walsh has had to refine new polishing techniques. The polishing process must remove material from the surface of the mirror very slowly and in a very controlled fashion. Sapphire also behaves differently from glass when suspended. Glass sags under its own weight, changing the shape of the polished surface; sapphire does this to a lesser extent.

While a mirror may have a perfect shape when lying down, when suspended it changes shape like a sweater on a washing line! Walsh and his team have overcome these and other difficulties, and now have 'a unique capability world-wide'. In fact, the CSIRO opticians are now producing the mirrors for the US gravitational wave observatory, LIGO.

The super nature of the mirrors causes its own problems. The very intense light built up within the optical cavities can attract and trap particles of dust and then drive them into the mirror surfaces where they are baked, damaging the mirror coatings. In addition, the tiny amount of light absorbed by the mirror coating causes heating, which in turn causes the mirror to expand, ruining the precision curve of the mirror. These problems mean taking extreme precautions with cleanliness. Many people are investigating new materials for making mirrors which can conduct heat away more effectively. Sapphire looks very promising, but many people hope that the ideal material, artificial diamond, one day will be available in sufficient size and purity.

The next challenge of laser detectors is in stabilising the laser itself. Most lasers fluctuate in frequency by millions of hertz or more. Laser interferometer detectors require the laser to be stable to within thousandths of one hertz. That means the frequency fluctuations must be less than one part in 10^{18}, better than the best clocks in the world. The accuracy does not have to be maintained for long—only for 10 to 100 milliseconds—corresponding to the frequency range 10 to 100 hertz over which the detectors should have best sensitivity.

It is still not known how accurately lasers can be stabilised. The problem of frequency control boils down to one of length variations, because the laser frequency is usually controlled by 'locking it' to the frequency of an optical cavity. The mirrors are supported on vibration isolation mounts, so that they experience very low levels of vibration above about 10 hertz—say less than 10^{-19} metres. But the mirrors absorb a little bit of heat from the laser beam, and the intensity of the laser beam varies. This causes the mirror to expand, and it is very easy to show that just a tiny change in laser intensity is sufficient to change the mirror spacing by more than expected in a gravity wave. The problem isn't all that bad, however, since much of the thermal expansion in one mirror is cancelled out by the expansion of the other mirror—identical in nature and environment.

Even more subtle things can happen to interfere with the interferometer's sensitivity. If the laser wobbles a little bit, say because a sound in the room vibrates a mirror in the laser, the beam strikes the mirror in the interferometer at slightly different places. This would not matter if the mirror was perfect. But nothing is. The light absorption of the mirror will vary over its surface, and therefore vary as the beam vibrates. This changes the laser intensity which in turn changes the length of the cavity by varying the temperature of the mirror. In other words, sound in a room can create an apparent gravitational wave.

These examples show how even the most subtle phenomena become catastrophic noise sources at the extraordinary sensitivity needed by the laser interferometer. Laser interferometer researchers working on gravitational waves spent most of the 1980s investigating all of the things that can go wrong with their detectors. Every possible flaw, every possible chain of events was studied. Eventually, the researchers convinced themselves that they knew how to build a full-scale working detector and asked their respective governments for the money.

Herein lies another saga: the struggle for money. Let's look at just one story as an example. In 1988 the Glasgow group was offered about £4 million towards a detector if they had an international partner. They asked the Australians if they would be interested in joining forces. This offer had a major effect in Australia: it made us realise that we could participate in large science projects and that a project of this sort was perfectly matched to Australian skills and resources. Much more, we quickly realised that a laser interferometer could bring enormous benefits to technology and industry in Australia. To be part of the unfolding and development of new laser technologies, new optical technologies, new vacuum technologies, new measurement systems and ultimately new astronomy could only be beneficial to the underfunded and ageing physics community in our country. Meanwhile the British funding offer was tied to the European Union, and Jim Hough's group planned to collaborate with Germany. Then outrageously, the British Science and Engineering Research Council withdrew the offer for funding of the British team. Then Germany reunified and the German group lost its expected funding. Fortunately, gravitational wave physicists are very resilient. Progress is always slow, experimental setbacks are a part of life, a political setback is just another obstacle to be overcome.

CHAPTER 13
NEW DEVELOPMENTS, FUTURE TRENDS

The dark cloud that has hung over the funding of large gravitational wave observatories at least has a silver lining. While physicists waited for the money, new ideas and technologies emerged which allow improved detector sensitivity, simplified operation and, eventually, high sensitivity in smaller-scale instruments. Let's take a look at a few of these improvements.

The most exciting new idea came from a brilliant young physicist with the Glasgow group called Brian Meers. He was a keen mountain climber and a fine drinking companion. His suggestion was similar to Ron Drever's earlier idea of recycling the laser beam power that we looked in Chapter 12. Meers's idea was just as simple and just as surprising. Whereas Drever suggested recycling the laser beam using a highly reflective mirror in the laser input beam, hence the name power recycling, Meers suggested putting another mirror on the interferometer output so that the signal was sent back into the instrument, hence recycling the signal. When you first think about it, it seems unlikely that such an arrangement would work: surely the signal would never make it out of the instrument for you to see it? What happens, however, is that the signal recycling mirror allows the signal to build up in the interferometer in a narrow range of frequencies, similar to the way the mechanical resonance of a bar allows it to gain sensitivity and reduce the noise. The other way it helps is that light escaping from the interferometer due to the imperfect shape of the mirrors is reinjected, and the effects of the imperfections are cancelled out, a process called 'wavefront healing'.

Initially, experimenters thought that interferometers would be built with optical cavities in their arms, plus a pair of recycling mirrors. This becomes complicated because the position of each mirror has to be controlled to an exquisite accuracy. Every extra mirror is an enormous additional complication. Several groups are now testing much simpler interferometers, where all the resonant gain is achieved by using very high performance signal and power recycling without any extra sensitivity-enhancing elements in the arms (i.e. no delay lines and no cavities). Very sadly, Brian Meers was killed in a mountaineering accident soon after his elegant invention.

We now realise that a recycling interferometer looks like a simple optical cavity as seen by the laser, and like another optical cavity as seen by the photodetector. There is no need for extra optical cavities. The only problem is that, if substantial resonant gain is to be achieved, the beam splitter which divides the light must have extremely low losses. This presents a major challenge.

The lasers used in all the successful prototype laser interferometer projects were argon ion lasers. These devices are better thought of as very powerful electric water heaters which produce a little bit of light on the side! A very powerful gas discharge through argon gas creates the conditions for lasing, but it is a most inefficient process. Most of the power is converted to heat, which is carried away by a water jacket. You get perhaps 20 to 50 kilowatts of hot water and only a few watts of light! The inefficiency and heat of the gas discharge also ensures that argon lasers are very noisy: the light intensity and light frequency fluctuate wildly.

Researchers have been able to tame these beasts, but only with tremendous effort. In the meantime, a beautiful new laser has come along: the diode pumped neodymium-YAG laser. Despite the mouthful of a name, this device is very high-tech, wonderfully simple and efficient. You start with a little artificial crystal of garnet—yttrium aluminium garnet to be precise—to which has been added a small amount of neodymium impurity, hence the name neodymium-YAG.

A semiconductor diode laser, just like the ones found in CD-players, is used to shine laser light (at a wavelength of 0.8 micrometres) onto the crystal. The diode laser is another beautiful piece of technology but, to the user, is just a tiny plastic-encapsulated gadget with two battery wires attached. What comes out of the crystal is a pure, low-noise beam of infrared light at a wavelength of 1.06 micrometres. Add more of these diode lasers and the output beam becomes brighter and brighter. The little crystal adds up all the low-quality diode laser light and converts it into a very high-quality infrared light, and this can be very powerful.

Being infrared, the laser light is invisible, which is a bit annoying—it is much more convenient to be able to see laser beams directly—but is easily visible to infrared-sensitive video cameras. The frequency of the light can be doubled (although not yet reliably) which makes it appear green, almost the same colour as the argon laser.

The efficiency of these devices makes them extraordinarily useful for all sorts of applications, and also very scary 'ray guns'. High power versions could easily be used to cut anything from masonry to clothing. The day is coming when buildings could be demolished by cutting them into neat little blocks and carrying them away piecemeal. The same technology can be used to fell trees and cut them into timber, harvest crops and butcher meat, all cleanly and quickly. By the same token, their potential as weapons is terrifying. At present they are too costly for most of these applications.

A critical problem for all gravitational wave detectors is isolation from local sources of vibration. In a typical city, the ground is constantly vibrating up and down about a micrometre at a frequency of about 1 hertz. At higher frequencies, the amplitude becomes less, usually inversely with the square of the frequency, so that 1 micrometre at 1 hertz becomes 0.01 micrometres at 10 hertz, one-tenth of a nanometre at 100 hertz and so on. The vibration level is so low we can't hear it, but even at 1000 hertz the size of the vibration is 100 million times larger than the gravitational waves we want to detect. Between about 0.1 and 0.2 hertz there is a much larger vibration due to ocean waves breaking on the shore. These waves produce long lines of surface waves which travel across the countryside at a speed of about 300 metres per second, and are detectable hundreds of kilometres inland.

The isolation of vibrations is a common problem in modern technology. Almost every device with a motor or engine has some form of rudimentary vibration isolator. In a car, for instance, you are isolated from vibration due to the road by springs and shock absorbers. The springs deflect whenever you drive over a pot-hole or speed hump, while the shock absorber, nothing more than an oil-filled cylinder with a loose-fitting piston that forces its way through the oil, stops the car from bouncing up and down like a yo-yo after encountering the pot-hole.

The same principles are used in gravitational wave detectors, although they have to be much more elaborate to be good enough for a gravitational wave antenna. Following the car analogy, the vibration isolation for a gravitational wave antenna should be so good that if it was on a car you could hit a speed hump at 200 kilometres an hour and the inside of the car would move no more than the size of the nucleus of an atom.

How can you achieve such superb performance? The answer, as we saw in Chapter 11, is surprisingly simple: mount many isolators on top of one another. Of course, you also need to make sure the isolator doesn't cause any vibrations of its own. In a car's shock absorber, for example, the noise of the flowing oil and the sliding surfaces would create unacceptable levels of vibration that would swamp all the isolation.

The very first gravity wave vibration isolators were simple: many layers of lead or steel and rubber. Some types of stiff rubber combine elasticity with large energy absorption. Such rubbers have been designed especially for vibration isolation. The next step in vibration isolation design was to try to use very-low-loss mass-spring systems—pendulums for horizontal isolation and springs for vertical isolation.

On their own, cascaded mass-spring systems have the same problem that a car has without shock absorbers: it will spring up and down for a long time after you hit the bump, and successive bumps can make it build up huge resonant vibrations. The solution to this is to replace the shock absorber with an electronic system that senses vibrations and feeds back forces to suppress them. This is called cold damping. A good example is a child's swing. If you push a child on a swing you can push in phase—that is push as the child starts to move away from you—and build up the swing, or you can actively damp out the swing by pushing out of phase—that is, push when the child is moving towards you—and quickly stop the swinging. The cold damping process is noiseless as long as you can sense the motion perfectly and apply forces without adding noise. In reality, of course, there is no such thing as a perfect measurement, but cold damping can operate extremely well.

The Perth group have developed vibration isolators using this principle. They work so well that for all frequencies above about 20 hertz they have been unable to detect any vibration at all at the limit of the detector sensitivity. Even when the support structure is deliberately shaken, no vibrations are detected.

This is not the end of the story, however. Every vibration isolator has a minimum frequency of operation. In a car you probably don't want to follow the bumps, but you do want to follow the rises and falls of the road. The frequency of a car's suspension is set to give the optimum balance between following hills and not following bumps. Somewhere in between the two, just at the cut-off frequency, you usually have the worst response; hence the effectiveness of speed humps.

For gravitational wave antennas, the cut-off frequency is similar to that of a car suspension, a few hertz. This is inconvenient because at this frequency the ground vibrates a lot, so that the suspended elements in a laser interferometer—mirrors and beam splitter—jangle around a lot even though at higher frequencies they are quite stable. Yet to operate a laser interferometer the components must be extraordinarily well aligned. Motions must be much less than about 10^{-10} metres, just to allow operation. In all devices operated so far, this has been achieved by using intricate and complex servo-control systems to steady everything as much as possible.

It would be much more convenient if the vibration isolation cut-off frequency was much lower, say one cycle per minute. At this frequency, the mirrors would take a minute to complete one cycle of oscillation. Above this frequency all vibrations would be cut off. Even the vibrations due to ocean waves would be greatly reduced. It would make everything so slow and steady that operation would be very simple and easy.

You could achieve this if you could suspend the mirrors and beam splitter from pendulums a kilometre long. They would then take 60 seconds to complete one swing, and only the most minute control forces would be needed to steer them into position. It sounds great, but even theoretical physicists famous for their crazy ideas would agree that a 1-kilometre-long pendulum is rather impractical.

Many people looked for a solution to the problem: was it possible to create a pendulum that behaved as if it were a kilometre long and yet was small enough to be built and used? The Perth group finally came up with the solution, a solution so beautiful that they thought about patenting it. About a year after they published their results, however, they found out that their idea had already been invented and implemented—by James Watt, the great steam engineer! Watt's linkage, as it is called, came from the need to couple hinged objects to a piston which moves in a straight line. Watt apparently was more impressed with this invention than of his steam engine, so at least the Perth group didn't feel too ashamed of their late rediscovery of the super-pendulum.

The Perth system consists of a platform suspended from two hinged arms, one above, one below. The key is the near straight-line motion of the mass, because this mimics the near straight-line motion of the end of a 1-kilometre-long pendulum. The prototype device stood a mere 20 centimetres high, yet behaved the same as a pendulum a kilometre long, reducing vibrations at 10 hertz by a factor of 100 000.

Groups all over the world are trying to outdo each other with better and cleverer techniques. We can feel confident that there are no insurmountable difficulties in making vibration isolators and that vibration isolation will not be a problem for gravitational wave observatories.

Laser interferometer gravitational wave observatories are difficult to build on Earth because of the constant battle for good vacuums and vibration isolation. There is yet another problem facing gravitational wave detectors: gravity gradient noise. This is the effect of direct gravitational forces on the test masses from moving masses nearby. Everything from people and machines to changes in gravity due to changes in temperature of the surrounding air can produce gravitational forces that might mimic the passing of a gravitational wave. For example, when a cloud passes over a 3-kilometre laser interferometer, it causes the buildings and the surrounding air to cool and contract. The gravity changes from such phenomena happen over a period of seconds, relatively slowly but big enough to swamp a gravitational wave signal. So on the time scales corresponding to frequencies below about one hertz, gravitational wave detectors on Earth are impossible to build.

There's a simple solution: leave Earth and head for space. In space you don't need a vacuum pump and there is no limit to the length of your detector. Interestingly, you do need some form of vibration isolation because the stream of particles arriving from the Sun, known as the solar wind, buffets the spacecraft minutely and constantly. From the point of view of gravitational wave detection, this minuscule effect would be disastrous.

Vibration isolation in space can be achieved by what is called 'drag-free technology'. If the test mass, such as a mirror for a laser interferometer, is placed in the middle of the spacecraft, it is shielded from the buffeting solar wind. Surrounding the mirror are sensors which indicate exactly where the spacecraft is, relative to the mirror. Anytime the solar wind causes the spacecraft to move, the spacing between the mirror and the spacecraft changes, which is detected by the sensors. Tiny thrusters on the outside of the spacecraft then fire to reposition the spacecraft so that it is once again centred on the mirror.

Space-based gravitational wave detectors were first proposed by Peter Bender at the Joint Institute for Laboratory Astrophysics in Boulder, Colorado. He worked for years on the design, taking into account the enormous number of factors that could make it go wrong. The idea was considered by NASA, but later abandoned.

The European Space Agency has since picked up the project, with the backing of gravitational wave physicists across Europe. Bender's gravitational wave detector now has a name: LISA, for Laser Interferometer Space Antenna. It is hoped that LISA will fly between 2010 and 2020.

LISA consists of four spacecraft using drag-free technology to create a laser interferometer. The spacecraft will occupy a stable solar orbit just behind the Earth, each one 5 million kilometres from its nearest companion. Two central spacecraft transmit separate laser beams out to the distant spacecraft, which then amplifies the laser before returning it to the central spacecraft. The returning signal at one of the central spacecraft is retransmitted to the other central spacecraft. There, the two return signals are recombined. The interference between the two beams is monitored and transmitted to the Earth. The signals tell the researchers about the relative spacing of the two spacecraft. As the spacecraft slowly orbit the Sun they drift slightly, causing the interference signal to vary between bright and dark light. But this orbital motion is relatively easy to calculate, as are its effects on the signals. On Earth, the orbital motion effects can be subtracted from the signals. What is left may be the signatures of gravitational waves.

Because the distance between the spacecraft will be so vast, it will take light 17 seconds to travel between any two. This means that gravitational waves with frequencies above one cycle every 17 seconds will be difficult to detect using LISA. Where the orbiting observatory will come into its own, however, is in the detection of low-frequency waves of one cycle per 100 or 1000 seconds. At these lower frequencies, LISA should be able to detect plenty of binary stars and neutron star pairs a few thousand years away from coalescence. LISA should also be able to hear the background rumbling of all the binary stars in the Galaxy.

Hubble Space Telescope images of very distant galaxies show an enormous amount of interaction. In the young Universe it looks as if the new-born galaxies collided frequently, perhaps creating the ancient giant elliptical and spiral galaxies we see in our own neighbourhood. Many of these nearby galaxies seem to have black holes at their centres. Some, like our own Milky Way, may have black holes of 'only' a million solar masses but others, like the giant elliptical galaxy M87 appears to have a black hole of 3 billion solar masses in its core.

Since black holes seem to power quasars, most astronomers believe that the black holes formed in the early Universe, starting as the source for quasars and going on to become the central black holes in the nearby galaxies. The nearby galaxies are the oldest objects we can see in the Universe because the light travel time from them to us is small, only a few million years. Thus, it seems likely that the black holes in our neighbourhood have also collided and interacted and merged sometime in the distant past. This means that the entire story of binary neutron stars may be repeated on a galactic scale, with a million or even a billion times more mass, and black holes instead of neutron stars. Such mergers would create gravitational waves with frequencies of from one cycle per day to one cycle per minute, right in the LISA band.

The strength of the gravitational waves from such events is such that any event in the visible universe could be detected with LISA. This means that space-based gravitational wave antennas can monitor the entire visible Universe for massive collisions or mergers of black holes. There are about one million million galaxies, most of them experiencing one or more collisions in the age of the Universe. When compared with the age of the Universe, there should be, on average, one merger every few days. If black holes occur in only 1 per cent of galaxies, however, mergers might be rarer than we think. Black hole mergers might be a weekly occurrence, or they might be quite rare. We can only tell by listening. If we do hear the distant rumble, we will have indisputable evidence for the existence of black holes and for the violent past from which our Universe has emerged. LISA is perfect for listening in to these intergalactic tsunamis.

CHAPTER 14
THE VISION OF GRAVITATIONAL WAVE ASTRONOMY

Gravitational wave astronomy will probably not begin in earnest until the twenty-first century. By then, detectors will have improved, and LISA may well have ventured into space to seek out low-frequency gravitational waves. The first signals will probably contain surprises, and later detectors will be optimised to resolve the first glimpses of the violent Universe. Throughout the next century, the gravitational wave spectrum will slowly unfold. As it does, the reality of reverberating and fluctuating space-time will become clearer and clearer. The unpredictability of space, of geometry itself, will become part of our cultural foundations.

Watching the ripples on a cosmic sea, we will witness the birth of black holes, image their vibrations, and perhaps understand the nature and significance of the singularity hidden in their infinitely dense cores. The singularity at the heart of a black hole may be similar to the singularity from which the Universe began. In gravitational waves we should be able to listen to the murmurs from the earliest moments of creation. With observation and theory, we may even grope our way to the point where we can answer not only how, but why the Universe began.

It has taken a century for the electromagnetic spectrum to be fully harnessed. Today the electromagnetic window to the Universe is wide open: giant radio and optical telescopes on Earth and orbiting observatories like the Hubble Space Telescope and the Einstein Gamma Ray Observatory allow us to view the Universe in electromagnetic radiation from 30 megahertz to almost a billion billion billion megahertz (10^{33} hertz)!

Today, gravitational wave detectors are at the place of internal combustion engines fifty years before the model-T Ford. It is a time of vigorous learning, enormous creativity, and progress which seems excruciatingly slow to impatient gravitational wave researchers. However, history will see this as a golden age of invention and exploration.

Almost 100 years ago, Professor W. H. Preece gave the following account of the momentous discoveries of electromagnetic waves in a speech to the Royal Institution in London:

> Science has conferred one great benefit to mankind. It has supplied us with a new sense. We can now see the invisible, hear the inaudible, and feel the intangible. We know that the Universe is filled with a homogenous continuous elastic medium which transmits heat, light, electricity and other forms of energy from one point of space to another without loss. The discovery of the real existence of this 'ether' is one of the great scientific events of the Victorian era. Its character and mechanism are not yet known to us. All attempts to 'invent' a perfect ether have proved beyond the mental powers of the highest intellects. We can only say with Lord Salisbury that the ether is the nominative case to the verb 'to undulate'. We must be content with a knowledge of the fact that it was created in the beginning for the transmission of energy in all its forms, that it transmits these energies in definite waves and with a known velocity, that it is perfect of its kind, but that it still remains as inscrutable as gravity or life itself.

Preece's remarks were made just before Michelson and

Morley's proof of non-existence of the aether had been fully accepted. Replace the word 'ether' with 'space-time' and there is little to dispute, except that gravity is not quite the mystery it was in 1897. We now know that space-time can sustain two completely different types of 'undulations', the electromagnetic or hertzian waves of Preece's speech and the waves of geometry—the ripples in the fabric of space-time—known as gravitational waves.

The existence of gravitational waves has been proved and new detectors across the planet are listening intently for the first ripples of the cosmic sea. We have names for the different parts of the electromagnetic spectrum—radio, microwaves, light, X-rays and gamma rays. Most of the spectrum has been utilised to the benefit of humankind. In stark contrast, the gravitational wave window to the Universe has opened no more than a crack. We have no names for the different parts of the gravitational wave spectrum. In principle, gravitational waves could extend over the same enormous range of frequency as do electromagnetic waves, from as low as one cycle in the age of the Universe to as high as the highest-frequency electromagnetic waves. For frequencies above one megahertz we have little idea of the possible sources or the nature of the detectors. However this means very little. Eighty years ago we had no idea that stars produced electromagnetic waves outside the optical band. What awaits gravitational wave astronomy is completely unknown. Will we discover new paradigms, or only more details of what we can already see? Will gravitational waves teach us anything at all? There is only one thing for certain: we can learn only if we continue to search. The first detection of these waves will be the beginning of a new branch of astronomy.

EPILOGUE
'A GIFT OF WONDER'

'Come on, Grandpa,' the young girl called back to me, running along the beach to the rocky outcrop. She ran so close to the water-line that her footprints were wiped clean by the next wave. When she reached the rocks she stepped carefully into the rock pool and bent over, with her face just over the water, staring intently. Suddenly, she pounced. 'I've found a cowrie,' came the shrill voice. No doubt, before long she would have a collection of treasures.

This beach always brings back memories of the old man I met here so many years ago. The water in the rock pool still ripples whenever I nostalgically throw in a sea shell. I remember how he wanted me to understand the importance of the search for gravitational waves. They had been a mystery to me before, something only a physicist could understand. That encounter had changed my life. He had shown me that gravitational waves could be understood by everyone. He was right, despite how bizarre the ideas had seemed at first.

I climbed a few steps up the rocks to get a better view of my grand-daughter's treasure hunt. Her mother was about that size when gravitational waves were discovered. What a stir it had caused; what headlines! PHYSICISTS DISCOVER HOLY GRAIL. For a whole week every gravitational wave physicist was a celebrity. From some stories you could believe that next week our cars would be running on gravitational waves! Yet, I remember my earlier sense of impatience . . . year after year there had been no news of gravitational waves. I certainly started to doubt if they would ever be detected. That had made the discovery all the more dramatic, especially when none of the theorists had correctly predicted the waveform and nobody could explain exactly what caused those first signals. Now, with the space laser observatory and the lunar lab churning out signals and TV ads and pop musicians using them incessantly, one can almost forget that those weird sounds from space are gravitational waves.

I remember wondering, at the time of the announcement, whether the old man had heard of the discovery. Had he lived long enough to know that those ripples in space he had introduced to me had been discovered? I never saw him again. He disappeared, along with his footprints on the sand which had been washed away by the waves he had watched so intently. Yet he left something behind that would never be erased. Now it is hard to imagine why the idea of curved space had seemed so weird. And it is hard to imagine a universe where you cannot listen to gravitational waves. Yet only decades ago, humanity was stone-deaf and could not even imagine that space could transmit sounds.

Although it never made the same big headlines, the discovery of G-source was equally important . . . especially after all the cosmologists had been convinced that superstrings held all the answers. And the origin of . . .

'Grandpa, are you falling asleep again?' What an indignant voice. She revised her comment when she saw that I wasn't asleep.

'Well, what are you thinking about?' Her hands were full of shells and seaweed. I looked down into those questioning eyes. She always wanted to know what I was thinking, and everything else as well. What holds the Moon up? Why do shells have patterns? Why can't we see air? If only everyone could hang on to that priceless gift of wonder. Some do, while the rest of us need someone like the old man who years ago rekindled that spark of curiosity in me. I always answered her questions as best I could, and this time it was easy to answer.

'I was thinking about gravitational waves,' I said, smiling at her fondly.

'I know what gravity is,' she said triumphantly, but then frowned. 'How can you have waves of gravity?'

'Come over here,' I said, walking to the small rock pool. I stooped down to pick up a shell and held it over the water. 'Watch the water,' I told her, and then dropped the shell into the pool. 'Do you see those ripples . . .'

INDEX

acceleration 23, 24, 25–6, 28, 29, 36, 40–1
Adams, John Couch 16, 17
Airey, George Biddell 16, 17

Badde, Walter 80
Baker, Don 87
Basov, Nicolay 117
Bell, Alexander 8
Bell, Joycelyn 83
Bender, Peter 169–70
Big Bang 111, 114, 115
Billings, Professor 148
black holes 63, 64, 66, 81, 82, 83, 93, 104–15, 173
Blair, David 141
Bondi, Herman 63
Braginsky, Vladimir 26, 126, 143–4, 145

Carter, Brandon 108
Chadwick, James 80
Challis, James 17
Chandrasekar, Subrahmanyan 80, 81, 98, 108
closed time-like loops 109–10
Colegate, Sterling 106
Copernicus 3
Crab Nebula 71
cryogenic resonant mass detectors 130, 134, 148–9
curvature 31–2, 33

Deutsch, David 110
Dicke, Robert 26
dipole oscillation 56
Doppler effect 52, 84, 89, 90, 91, 96
Drever, Ron 148, 149, 150, 154–5, 156, 162

Eddington, Arthur 41–2
Einstein, Albert 3, 4, 6, 10, 11–23, 29, 34, 35, 38, 39, 41, 42, 54, 62, 63, 79, 93–4, 105, 106, 107, 117
Einstein Gamma Ray Observatory 174
Einstein rings 50
electromagnetic waves 55, 56, 173, 175
electron degeneracy pressure 81–2, 105
Eotvos, Baron Lorand von 26, 27
Equivalence Principle 23, 26–7
Euclidean geometry 2, 3, 4, 5

Evans, Reverend Robert 73–4
Everitt, Francis 47

Fairbank, Bill 130–1, 134, 135–6, 138, 141, 142, 148
'fifth force' 27
Forward, Robert 148
frame dragging 47, 49
freefall trajectories 34, 43, 45

Galilei, Galileo, 2, 3, 12, 26, 28
Galle, Johann Gottfried 17, 21
gamma rays 43
Garwin, Richard 122–3
Gauss 3, 4
Gautama Buddha 128
geometry 2, 3, 4–6, 19, 32–3
Gertsenshtein, M. 147
Gibbons, Gary 123, 124
Giffard, Robin 126, 145
Global Positioning System (GPS) 45, 46
Gold, Thomas 84
gravitational:
  collapse 105–6; lensing 49–50, 51, 53–4; mass 23–4, 26, 27, 28, 117; time dilation 34–5, 43, 44, 45, 46, 79; waves 1–2, 8, 11, 23, 35, 55–64, 65, 67, 72, 76–7, 93–4, 95–103, 108, 111–13, 115, 116, 117–29, 130, 131, 133, 134–45, 147–61, 162, 165, 167, 168–9, 170–1, 173–5
gravitational wave detectors, cryogenic 134, 136
gravity 1, 4, 11;
  and acceleration 23, 24, 28, 29, 40–1; law of 13–16, 17, 18, 20, 23; strong 79
gyroscopes 47–9, 130–1

Hamilton, Bill 131, 134, 136, 150
Hawking, Stephen 108, 110, 111, 123, 124
Herschel, William 15
Hewish, Antony 83
Hirakawa, Professor 141
Ho Jung Paik 138, 139, 140
Hough, Jim 148, 154, 161
Hubble Space Telescope 53, 54, 113, 171, 174
Hulse, Russell 95, 99

inertia mass 24, 25, 26, 27, 28
inflation 114
interferometer 8–9, 147–8, 149, 150–3, 154, 155, 156–61, 162–4, 168–72
Israel, Werner 108

Kepler, Johannes 15, 19, 38
Kulkarni Shrinivas 87

Landau, Lev 80
laser 117
Laser Interferometer Gravitational Observatory (LIGO) 155, 157, 158, 159, 168
  see also interferometer
Laser Interferometer Space Antenna (LISA) 170, 171–2, 173
Leverrier, Jean-Joseph 16, 17, 21
light:
  curvature 40–3; speed of 8, 10, 40; waves 6, 8–9, 149

Lyne, Andrew 91

Manchester, Dick 86
mass:
gravitational 23–4, 26, 27, 28, 117; inertia 24, 25, 26, 27, 28
massive compact halo objects (MACHOs) 51
Meers, Brain 162
Mercury 17, 21–2, 29, 35, 38, 40, 95
Michelson, Albert 3, 7, 8, 9, 147, 174
Microwave Amplification by Stimulated Emission of Radiation (MASER) 117
Mitchell, John 104
Moon 13, 14, 18, 20, 24
Moore, Baud 157
Morley, Edward 3, 7, 8, 9, 147, 175
Morrison, Phillip 122, 123
motion, laws of 12–13, 25, 35

National Science Foundation 154, 155, 156
Neptune 17
neutron stars 63, 64, 71, 79–94, 105, 106, 156
Newton, Isaac 2, 11–22, 23, 25, 28, 29, 32, 35, 39, 41, 54, 125
noise 118–20, 144–5

Onnes, Kammering 131–2
Oppenheimer, Robert 80, 83, 105, 106

Pacini, Franco 83
Pauli Exclusion Principle 81
Pennypacker, Carl 74
Penrose, Roger 107, 110
piezelectric crystals 118, 119, 125, 148–9
Pizzella, Guido 127–8, 135, 148
photons 43–4
physics 2, 11–12
Planck 144
planetar orbits 13–22, 38–40
polarisation 58, 60
Preece, Professor W.H. 174
precession 40
Prokhorov, Aleksandr 117
Ptolemy 2, 3, 6
pulsars 79, 83–94, 95–103, 104, 115
Pustovoit, V.I. 147

quantum mechanics 138, 143, 144–5, 151
quantum non-demolition (QND) 145
quasars 50, 52–3

relativists 46, 63
relativity, general 11, 18, 22, 23–37, 38, 40, 41–2, 43, 44, 49, 54, 79, 107, 116
resonant mass detectors 130, 131, 134–44, 147, 148–9, 150
ripples in curved space 6, 54
Rutherford, Ernest 80

Schwartzchild, Karl 105, 106, 107
singularity 104, 106, 107, 110
Snyder, Hartland 105, 106
space 1, 2, 11, 17, 34; curved 4, 6, 10, 18, 19–20, 33, 35, 39–40, 41, 42–3, 49, 56, 115; waves 10, 11, 55
space-time 1, 2, 17, 18, 19,

20, 23, 28–30, 32, 33, 34, 35, 36, 41, 42–3, 47, 49, 51, 55, 60–1, 114, 115; ripples 54, 56, 62, 114, 115
'superconducting quantum interference device' (SQUID) 48–9, 139, 140
superconductivity 130, 131–3, 135–7, 138, 139
supernova 67–78, 115, 126, 142

Taylor, Joe 95, 99
Thorne, Kip 154
time 1, 2, 17, 34–5, 43, 79
  see also closed time-like loops: gravitational, time dilation
Townes, Charles 117
triangles 4, 6, 31–2, 33
Tyson, Tony 122

Uranus 15–16

Vessot, Robert 45
Vogt, Robbie 155–6, 157
Volkoff, George 80, 83

Walsh, Chris 158–9
Watt, James 168
wave speed 7
Weber, Joseph 116–25, 127, 128, 129, 130, 134, 135, 138, 141, 148, 150
Weiss, Rainer 148, 149–50, 154
Wheeler, John Archibald 105
white dwarfs 81–2, 83, 85, 92, 105

Zwicky, Fritz 80